THIRTY BELOW ON CHRISTMAS EVE

INTERVIEWS WITH NORTHWEST OHIO VETERANS OF THE KOREAN WAR

COMPILED AND EDITED BY
ANDREW "BUD" FISHER

The University of Toledo Press

http://www.utoledopress.com

Thirty Below on Christmas Eve

Book design by Joan Bishop, The University of Toledo
Project assistance by Caitlin McCallum

ISBN 978-0-932259-15-8

The cover graphic combines a photograph of two of the 19 statues at the
The Korean War Veterans Memorial on the National Mall in Washington D.C.
with an image of a South Korean landscape in winter.

The publication of this book was supported in part by donations
from the following organizations:

Chapter 35 — Vietnam Veterans of America: "All gave some; some gave all."

The University of Toledo Reserve Officer Training Corp (ROTC)
"Rockets — Proud to Be!"

The University of Toledo Military Service Center

DEDICATION

The University of Toledo and the Library of Congress are partners in the Veterans History Project, a national effort to collect and preserve the memories of those men and women who have served their country.

There are thousands of volunteers across the nation who visit the homes of veterans to record there military experiences. Presently, the Library of Congress has an archive of 75,000 military histories. The University of Toledo has an archive of more than 500 recorded interviews of northwest Ohio and southeast Michigan veterans. Each of the 500-plus veterans interviewed also has a copy of his or her recorded interview.

This effort, ongoing for more than eight years, lead to my compilation of eighty World War II veterans' interviews, *What A Time It Was*, introduced on Veterans Day, 2009. The title refers to the years from the Great Depression of 1929 to the end of WWII, in 1945. The young men and women who persevered through those trying years are the book's subjects.

The positive reception that *What A Time It Was* received inspired a book about the Korean War veterans, *30 Below on Christmas Eve*. It, too, is a compilation of the words of the area's men and women who served their country.

The Veterans History Project, *What A Time It Was*, and now *30 Below on Christmas Eve*, all do honor to the men and women from northwest Ohio and southeast Michigan who answered the call to duty.
— Bud Fisher

TABLE OF CONTENTS

Appendices

To avoid re-defining the terms, battles, planes, and other references in *Thirty Below on Christmas Eve*, such words are starred (*) with the definitions given on page 124.

FOREWORD

by Howard McCord
Aviation Machinist Mate Third Class, U.S. Navy, 1951-1954

My first view of Korea was on a January dawn from the catwalk alongside the flight deck of the USS Essex. We were close enough to shore to see the hills clearly—black, rugged, with a heavy scattering of snowfields. The wind was colder than any wind I had ever felt before. I was wearing my foul weather jacket and my woolen watch cap, and they seemed to provide no protection from the cold. I thought of the soldiers and Marines in those hills, some perhaps trying to sleep in poorly heated tents, or on duty, or lining up for chow, but never really out of the cold. I watched for a few moments, and then headed down the ladder to a hot breakfast in a warm messing compartment. A big ship seemed like a very good place to be in this part of the world on that day.

The war had begun just days after I graduated from high school in 1950, and I spent the summer taking a course in analytical geometry, climbing desert mountains, and reading about the war in the newspapers. Fall semester started as our forces had been pushed down to Pusan. Then came the Inchon landing, and the tide seemed to turn. I took a full course load, shot for the ROTC Rifle Team, and kept up with the news from Korea. After the Chinese armies poured over the Yalu in late October, and cold weather began, the situation turned desperate. I decided to enlist in the Navy that December. I was called up on March 6, 1951.

My father served in the Navy in the Pacific in World War II, and my great-grandfather has been in the Union Navy, so it seemed the service to join. Years later, my son Colman would serve three tours on the DMZ with the 31st and 20th Infantry Regiments. He has his own tales of winters there, as well as of a hitch in the Navy.

Everyone's story in *Thirty Below on Christmas Eve* is different, but they were all far away from home in a dangerous environment under difficult conditions. People were trying to kill them, and the weather was often doing the same. The personalities that emerge from these interviews are those of strong, tough, dependable, levelheaded men who did not panic, but endured, and survived. Several had lived hard-scrabble lives in their youth, and had strong memories of the hardships of the Depression which I think helped them get through their ordeals.

You will meet Robert Lempke, who went into the National Guard at age fifteen, fought at Chosin, was captured by North Koreans for ten days, then freed by Marines, and returned home at the end of his enlistment, a combat veteran by age eighteen. Joel C. Davis was so moved by the plight of the children he saw that he started an orphanage, which grew well beyond his individual efforts as others joined in to help. Several went into battle in the early months of the war with inadequate equipment—World War II boots, no winter coats, and insufficient ammunition. Terrence Mohler's artillery unit was limited to three 155mm shells per day per gun, and as a radioman, he had no radio. Leo Barlow, a Marine and one of the Frozen Chosin, went from September till April without a hot shower. His feet were frozen, but he marched out.

Donald Griffith was captured at Chosin and was a prisoner for thirty-three months. When he escaped and was recaptured, his boots were taken and he was thrown into a pigsty for a month. He was also the victim of medical experiments, as were other prisoners. Robert Darr's story brought back particular memories to me. He was a driver at Operation Little Switch, when the more seriously wounded prisoners were exchanged. One of them was my friend John Chancellor, whose back had been broken. John recovered and was a fellow student with me at Texas Western College in 1956.

Andrew "Bud" Fisher has done an outstanding job in compiling these fascinating accounts and framing them with extremely useful historical information about the war and the major personalities involved in its conduct. If Korea is still fresh in my mind after sixty years, I know most people's memories don't include it. That's why such books as *Thirty Below on Christmas Eve* are so important, and why so many Korean veterans take the time to visit schools and tell the students about the war. The teachers are too young to remember it also. No textbook can be as vivid as a personal account of events lived through and witnessed.

Howard McCord directed the Creative Writing Program at Bowling Green State University for many years. A native of west Texas, his novella, The Man Who Walked to the Moon, *is a cult classic. McCord's* Collected Poems *were published by Bloody Twin Press in 2002. He lives in Wood County, Ohio.*

INTRODUCTION

To understand the origins of America's involvement in the Korean War, it is necessary to step back to the Yalta conference of February, 1945, where Joseph Stalin, Franklin Delano Roosevelt, and Winston Churchill—representing the Union of Soviet Socialist Republics, the United States, and Great Britain—agreed on the rebuilding of Europe. One important topic was the issue of free elections for all the newly liberated nations.

In spite of Stalin's agreement that free elections would be held in the territories under Soviet postwar control, it soon became obvious that puppet governments would be installed in the eastern European countries the Soviets occupied. This would have worldwide repercussions, affecting Asian nations as well.

President Harry Truman determined that the United States must act to save the free world from total communist domination. In March, 1947, he announced a policy of containment known as the Truman Doctrine—"the policy of the United States to support free people who are resisting attempted subjugation by armed minorities or by outside pressures." While the Truman Doctrine originally established U.S. support for Greece and Turkey, it became the basis of American policy during the period known as the Cold War (1946-1991), the political, ideological, and economic battle between Western nations and those aligned with the USSR for influence and strategic gains.

In June, 1945, the United Nations had replaced the largely ineffectual League of Nations, in existence since the end of World War I. The League had not been able to contain the Axis powers prior to World War II, but the UN was determined to be a proactive guardian of world peace.

Like the USSR, China had been a WWII ally of the United States. The Allies supported its war with Japan, bringing in supplies on the Burma Road until Burma was captured by the Japanese in 1942. During WWII, Allied planes flew over the Himalaya Mountains, the "Hump," to support China's war effort. Internal fighting had been ongoing between the Chinese government forces and the Chinese Communists since 1927, but was interrupted by WWII and the Japanese occupation. At war's end, the

Chinese civil war began again, until the Communists under Mao Zedong seized control in 1949. The Nationalist government retreated to Taiwan and the U.S.-China alliance ended.

Korea had been occupied by Japan since 1910. At the end of WWII, Japanese forces surrendered to the Russian forces occupying Korea north of the 38th parallel*, and to the American forces occupying the south. In 1947, the UN declared elections should be held throughout Korea, in order to choose one government for the entire peninsula. The Russians, however, refused to allow elections in their sector. In May, 1948, South Koreans elected a national assembly; in September, the North Korean Communists established the Democratic People's Republic of Korea. Both Koreas claimed the entire country, leading to border clashes from 1948-50. The Russians left in November, 1948, followed by the U.S. in 1949. Early in 1950, the U.S. indicated that the two Koreas were outside what the U.S. considered defensible Asian areas. On June 25, 1950, North Korean forces invaded South Korea, in what would become the first armed conflict of the Cold War.

The UN, declaring North Korea's action to be an invasion of international peace, demanded the North's withdrawal. When fighting continued, UN member nations voted to send military aid to South Korea. Sixteen countries sent troops and forty-one provided military equipment, food, or other supplies. U.S. General Douglas MacArthur was named commander in chief of the UN Command and directed the war effort from Tokyo, Japan. Despite Truman's commitment to an American policy of containment, the president agreed to American involvement.

In addition to South Korea, members of the United Nations force included the United States, Australia, Belgium, Canada, Colombia, Ethiopia, France, Greece, Luxembourg, the Netherlands, New Zealand, the Philippines, South Africa, Thailand, Turkey, and the United Kingdom. Supporting the Democratic People's Republic of Korea were the People's Republic of China and the Union of Soviet Socialist Republics, whose participation was both secretive and limited. Some ninety percent of the soldiers and supplies came from the U.S.

The North Koreans swept down the Korean peninsula, leaving the Allies controlling only a small area surrounding Pusan*, a coastal city on the southeastern tip of South Korea. The UN forces managed to hold a perimeter around Pusan, stopping the North Korean advance on September 8, 1950. Then on September 15, MacArthur launched a surprise amphibious landing at Inchon*, on the Korean west coast not far from Seoul, the South's capital, and the 38th parallel. The UN forces moved north rapidly and by mid-October were approaching the Yalu River*, the border between North Korea and China.

The Chinese, alarmed at the threat to their border and the not-so-secret Allied discussions of invasion and possible use of nuclear weapons, issued several warnings against any further UN troop movements. MacArthur dismissed the Chinese statements, not believing that China would risk men and materiel to enter the conflict he planned—and expected—to be finished by Christmas. But the UN commanders were unaware that the Chinese had already sent troops into North Korea. On October 25, the People's Volunteer Army (PVA) attacked UN forces near the border. By the end of November, an estimated 300,000 Chinese troops had crossed into North Korea.

A series of attacks, retreats, and counter-attacks continued through November and December, with fierce battles in the mountainous areas around the Chosin Reservoir* and the Ch'ongch'on River*, during what was the coldest Korean winter in one hundred years. Official temperatures hit -35 degrees; many soldiers remember even colder readings as they suffered severe frostbite and other cold-weather injuries. Any hopes for a swift end to the conflict were snuffed out and by the end of November, UN troops were in retreat back to the port of Hungnam. Chinese and North Korean troops again captured Seoul. By March, 1951, UN troops were able to retake Seoul as both sides began to dig in along either side of the 38th parallel. Fighting continued during the next two years for control of strategic positions, but neither side made important gains during what is known as the Stalemate Period*.

The full extent of China's participation in the Korean War is difficult to pin down. The Chinese People's Volunteer Army (PVA) was part of the Chinese People's Liberation Army, but had been set up separately to avoid any official war with the United States, whose troops were part of the United Nations offensive. The Chinese attacked again in November, 1950, along the Ch'ongch'on River and the Chosin Reservoir, forcing the 8th Army and then the Marines to make the longest retreat in U.S. history.

Although the Chinese were able to recapture most North Korean territory, the Marines inflicted such heavy casualties that an estimated forty percent of the Chinese force in Korea was rendered militarily ineffective. The Chinese mounted three more campaigns but were never again able to maintain territory. The final Chinese campaign, the Spring Offensive, began with 700,000 troops attacking U.S. I Corps at the battles of Imjin River and Kapyong, but by June, 1951, the Chinese troops were surrounded; the division fell apart and men began to desert or were captured.

General MacArthur had disagreed with the pace and strategy of the war, pushing for total victory, which included bombing Chinese bases in Manchuria, and other "all-out measures." When his statements became public, and were even read aloud on the floor of the U.S. House of Representatives, Truman felt he could no longer accept the general's actions. On April 11, 1951, the president and commander-in-chief replaced his world-famous and immensely popular general with Lieutenant General Matthew Ridgeway*.

In June, the Russians offered a cease-fire proposal; peace talks began in Kaesong, in southern North Korea, on July 10, 1951. The on-again, off-again negotiations continued through 1951, 1952, and into 1953, during which time some of the fiercest battles of the war were fought, including Heartbreak Ridge*, Old Baldy*, Porkchop Hill*, and the sieges of Outpost Harry*. But as in World War I, thousands of lives were lost but little or no territory was gained by either side.

Several prisoner exchanges were successfully completed in April and May, 1953. The armistice agreement, ending the fighting, was signed on July 27, 1953. A Demilitarized Zone* was established along the 38th parallel to separate the two countries, 2.5 miles wide along the final battle line. Almost all of the Korean peninsula had been damaged by fighting or bombing; property damage was estimated at $1 billion.

Previous casualty figures for the United States had been set at some 54,000. That death toll, however, included all service members who died while on active duty for any reason, not just those killed in battle. In 1993, the Department of Defense changed its reporting procedures, citing 33,686 battle deaths and 2,830 non-battle deaths in Korea, along with 8,186 missing in action. During the war period, 17,730 other military deaths were reported. The South Koreans suffered 138,000 military casualties.

The figures for North Korean and Chinese deaths are more difficult to determine. Most Western sources put the number of Chinese soldiers killed at around 400,000 while the Chinese themselves report 183,000 wartime deaths. Western estimates put North Korean deaths at around 215,000, while Chinese sources raise that to 290,000 killed. Around 3,870 soldiers from the United Kingdom, Canada, Australia, Turkey, France, Greece, Thailand, the Netherlands, the Phillipines, Belgium, New Zealand, South Africa, and Luxumbourg (2) were also killed. The numbers of wounded include 450,742 South Korean troops; 92,134 U.S. soldiers; 303,000 North Korean troops; 486,000 Chinese (Western estimate); and about 7,330 other UN troops. The total number of dead and wounded ranges from 1.2 to 1.6 million persons.

Civilian casualty numbers vary but median estimates place the total of North and South Korean civilians killed at 1,595,000. The three years of the Korean War resulted in the probable loss of some 3,000,000 lives.

As of this book's publication, no permanent peace treaty, officially ending the Korean War, has yet been signed.

— Bud Fisher and Molly Schiever

THE VETERANS HISTORY PROJECT

The University of Toledo and the Library of Congress are partners in the Veterans History Project, a national effort to collect and preserve the memories of those men and women who have served their country in the armed forces.

There are thousands of volunteers nationwide who visit the homes of veterans to record their military experiences. The Library of Congress currently has an archive of 75,000 military histories. The University of Toledo has an archive of more than 500 recorded interviews of northwest Ohio and southeast Michigan veterans. Each of the 500 veterans interviewed also has a copy of his or her recorded interview.

This effort, ongoing for more than eight years, led to *What A Time It Was*, a compilation of eighty World War II veteran interviews. The title refers to those years from the Great Depression of 1929 to 1945, the end of WWII. As the Great Depression was finally ending, the United States was faced with yet another worldwide crisis. The young men and women, who persevered through those trying times, are the subjects of the book, which was introduced on Veterans Day, 2009.

The positive reception *What A Time It Was* received was the inspiration for *Thirty Below on Christmas Eve*, interviews of Korean War veterans. The Veterans History Project, *What A Time It Was*, and now, *Thirty Below on Christmas Eve*, all honor the men and women of northwest Ohio and southeast Michigan who answered the call to duty.

KOREAN WAR TIMELINE

1945
8 August	Russian troops invade Manchuria and Korea
9 September	Japanese troops surrender to U.S. forces in South Korea

1947
11 November	UN forces leave Korea

1948
8 February	North Korean army activated
8 April	Truman orders U.S. troops to leave Korea
15 August	Republic of Korea (South Korea) is established
9 September	Democratic People's Republic of Korea (North Korea) is established

1949
29 June	Last U.S. troops leave Korea

1950
25 June	**North Korea invades South Korea**
29 June	Seoul, capital of South Korea, falls
30 June	U.S. troops to Korea
7 July	UN forces in Korea, under General Douglas MacArthur, Commander, Far East Command (FECOM)
15 September	U.S. Marines land at Inchon
27 September	UN forces in North Korea
14 October	China enters the war
19 October	Pyongyang, capital of North Korea, captured
27 November	Chinese attack at Chosin Reservoir
30 December	Russian MiG jets enter the war

1951

3 January	Communists again capture Seoul
18 March	UN forces retake Seoul
1 April	General MacArthur fired by President Truman
	General Ridgeway new commander of FECOM
12 June	General Mark Clark new commander of FECOM
23 June	USSR proposes cease-fire
10 July	Peace talks begin at Kaesong
17 August	Battle of Bloody Ridge begins
23 August	Russians end talks
5 September	Battle of Heartbreak Ridge begins
25 October	Peace talks resume

1952

29 August	1,400 bombers hit Pyongyang
8 October	Peace talks break off

1953

30 February	Peace talks resume
20 April	Exchange of sick and wounded POWs begins
18 May	Peace talks break off
10 July	Peace talks resume
27 July	Cease-fire signed

KOREAN WAR MAPS

Korean Peninsula

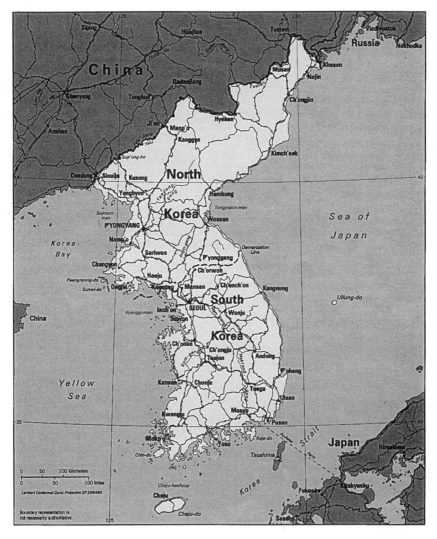

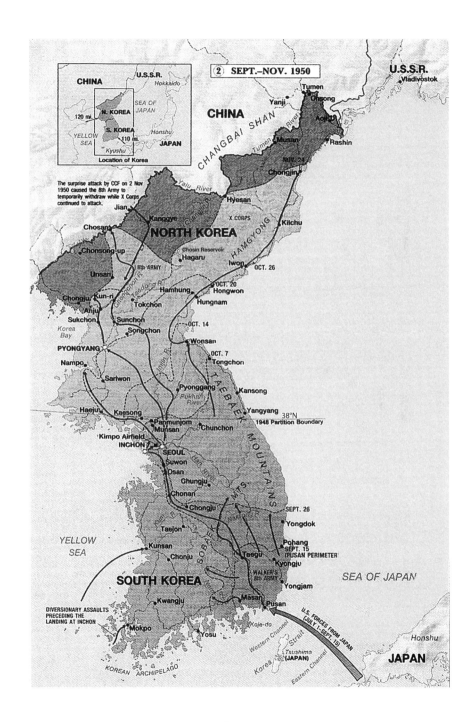

U.S. AIR FORCE

In 1907, the United States created the Aeronautical Division, an aircraft fighting group and part of the Signal Corps, renamed the Aviation Section in 1914. In May, 1918, it became the U.S. Army Air Service; then in 1926, the service was renamed the U.S. Army Air Corps. In June, 1941, the Air Corps became the U.S. Army Air Forces. In September, 1947, the Army Air Forces were abolished and all personnel were transferred to the new Department of the Air Force, a separate branch within the newly created Department of Defense.

B-29 dropping bombs on Korea.

Donald M. Callahan
U.S. Air Force

Donald Callahan didn't want to run up and down the mountains in Korea and he couldn't swim. That ruled out the Army, Marines, and the Navy, so he enlisted in the Air Force.

Where was your basic training?
Lackland Air Force Base, San Antonio, Texas.

How was basic training in Texas?
At twenty, I could handle almost anything. After basic, I went to Lowery Air Force Base, in Denver, for remote-control turret-system mechanics school and gunnery school. The B29 [Superfortress bomber, built by the Boeing Co., used in WWII] had a remote control system where the gunner was not actually in the turret firing the guns.

Where did you go from there?
Randolph Field, Texas [near San Antonio], to form up a crew.

What is a B29?
The B29 is a four-engine bomber that weighed 120,000 pounds, which was a big airplane in those days. It was the most sophisticated bomber in the Air Force before the B47 [Stratojet, built by the Boeing Co]. The B29 dropped the atomic bombs.

Where were you stationed in the East?
Kadena Air Base, on Okinawa, which was about 800 miles south of Korea or about two hours flying time. There wasn't much to do on Okinawa, but we could go on R&R* to the Philippines. We were in Japan just once, when we ran short of fuel and had to make an emergency landing, but we never got to see much because it was so foggy.

What was your mission in Korea?
Our job was to bomb bridges, factories, and air fields. There was a lot of

anti-aircraft fire and if there was no flak, there would be enemy planes. One night, we lost three planes out of nine, shot down by air-to-air missiles.

During World War II, air crews flew a certain number of missions, twenty-five or thirty-five, and then would return to the States. Did you have the same deal?
No, we had to stay for our regular tour of duty on Okinawa, which was six months.

Were you ever hit?
No, but we flew in single-file formation. Each plane followed the one ahead by three minutes and would drop its bombs. We had to stay in formation, even if there was flak.

When you left the Far East, you still had time left on your enlistment. Where did you go?
I went to Lake Charles [Air Force Base, later Chennault AFB] Louisiana. I was no longer part of a crew, but I flew my five hours to qualify for flight pay. When [it was] changed to a B47 base, we no longer flew. The B47 is a six-engine jet with the droop wings. The B47 had only one gunner, so I wasn't included. Instead, we were sent back to Randolph Field, Texas, to get a crew ready to fly a C97 [Stratofreighter, cargo plane, also built by Boeing] refueling plane. My enlistment was up before I flew the C97.

You were pretty well trained. Did you consider staying in?
I finally decided to take an early release and go home to Carey, Ohio. I used the GI Bill* to go to an aviation school and eventually went to work as an airplane mechanic for Owens-Illinois in Toledo until I retired.

Donald Callahan's decision to join the Air Force would shape the rest of his life. From 1953 until he retired, he enjoyed a lucrative career in aviation.

Wendell B. Lawrence
U.S. Air Force

*Wendell Lawrence always wanted to fly
and he wanted to defend his country, so he
joined the ROTC [Reserve Officers' Training
Corps] at Michigan State University,
graduated as a second lieutenant, and
committed to a four-year enlistment. He
then went into the United States Air Force.*

After you had taken your tests to qualify for the Air Force, where did you go?
I went to San Angelo, Texas, for basic flying training and then to Williams Air
Force Base, Arizona, for jet training.

Is a jet harder to fly than a prop plane?
Yes, but it's smoother and, of course, much faster.

How much training did you have before leaving for Korea?
We had one year of flight training and six months of aerial gunnery.

Do you think your enemy had the same degree of training?
Yes, when it was the Russian pilots, but not [for] North Korean or Chinese pilots.

What kind of plane did you fly in Korea?
We flew F84s*, mostly doing air-to-ground missions, but we would get hit by
Russian MiG15s*. The MiG was a faster, more maneuverable plane, but where we
had dual controls, they had single controls, so they were easier to damage and shoot
down.

Did you eventually have air supremacy?
Yes, when we got the F86*, we had a plane superior to the MiG.

Did you lose many planes to the MiGs?
No, we lost more planes to ground fire, because we were always flying so low for
air-to-ground missions. I was in a dogfight on my first mission, but managed to
outmaneuver him and he finally was low on fuel, so he left. We fought over Korea
and they came from China, so we had the fuel advantage.

Why were there Russian pilots in a war between the UN forces and North Korea?
The Russians were backing the North Koreans with planes, but the North Koreans

had almost no pilots and the Chinese had very few, so Russian pilots were used. The Chinese pilots were better than the North Koreans and the Russians were better than the Chinese.

How many missions did you fly in Korea?
One hundred.

Were you ever shot down?
I was hit once by ground fire, but managed to land safely.

If a pilot was shot down, did you stay with him?
We would circle him until a helicopter came to rescue him. When a plane was shot down and the pilot wasn't killed outright, he was usually rescued.

How long were you in Korea?
It took me five months to get in my one hundred missions, for rotation back to the States. I went to Korea as a second lieutenant and was promoted to first lieutenant before I left.

Where were you sent next?
I went back to the gunnery school, where I had trained, and became an instructor.

What happens at gunnery school?
We teach skip bombing, dive bombing, and aerial gunnery. I was there five years and then I was tapped to go to the Air Force Academy in Colorado as the baseball coach. I had played at MSU and a little pro ball before the service. I spent five years coaching and flying. Life at the academy was very good.

Then I was sent to Europe to fly F101s [McDonnell F-101 Voodoo] and F4s [McDonnell Douglas F-4 Phantom] with the NATO Forces. I went from Europe to Vietnam and was there for the Tet Offensive [January, 1968]. The North Vietnamese had no planes and the Russians and Chinese were not involved, so our greatest danger was the SAM [surface-to-air] missiles. I stayed in Vietnam for one hundred missions and was sent back home.

I had been in the service for twenty years. I could retire after twenty years, but I stayed another five. I could have stayed longer to make general, but that would have meant a desk job in Washington. I retired in 1976 at fifty years of age and returned to Toledo. I went to Bowling Green State University, got my master's degree, and began my second career, working for the Medical College of Ohio.

I have been all over the world and the U.S. is still the best place to be. It is the best country in the world, and I am happy that I was able to contribute my small bit to defend it.

Richard Piriczky
U.S. Air Force

Richard Piriczky and his good friend and schoolmate, Richard Wagner, were interviewed, one after the other, on April 18, 2006. Each commented during the other's interview, which made these interviews unique and interesting (Mr. Wagner's comments are in italic).

Mr. Piriczky wanted to enlist in the Marine Reserves at age sixteen, but his parents refused. At seventeen, he went to join the Army, but failed the physical. "They told me to go across the hall and try the Air Force," he said. The Air Force accepted him and off he went to the Lackland Air Force Base, San Antonio, Texas, for thirteen weeks.

How was basic training?

Lots of class work, marches, and close order drills. We had one twelve-hour pass to San Antonio, during our seventh week. At the end of basic, I sat with a master sergeant, who had a big catalogue of possible areas to specialize in. I said I wanted to be a gunner, but that was closed. I wanted to be a heavy equipment operator, but that was closed, too. Finally we found that the MPs [military police] were open, so I went to school at Camp Gordon, Georgia, for eight weeks, training to be an MP. After school, I went to Shepherd Air Force Base, Texas.

On June 25, [1950] I found out that there was a war in Korea. I didn't even know where it was, but two weeks later, I was headed there. There were 7,000 of us on a troopship, designed for 3,500. It was so crowded that the guys coming back from breakfast were meeting the guys going to supper.

We went to a British base in Japan first. The Korean War was a United Nations war, with about twenty nations taking part [seventeen combatant nations, five providing medical support]. We had the British and the Australians on the base. The Americans had by far the most people and the most equipment.

What kind of planes did you have?

The Royal Australian Air Force had the P51s* and the Gloster Meteor jets [built by the Gloster Aircraft Co.] and we had the B26* bombers.

So you were using jets at the time?
We were using the Shooting Star jet [F80/P80*], which was too fast for ground support. The old P51* was better, but because Korea was the first war where we used jets [the British and Germans had introduced jet aircraft towards the end of WWII], they got all the attention. Later, we got the F84* and F86*; they were better jets than the Russian MiGs*.

Were the Australians, flying the old P51, going up against the Russian MiGs?
Not too often, the P51s were used for ground support.

Mr. Wagner—I'd like to say something about the P51 pilots. They were the best pilots in the world. We used to say that you could tell if a P51 pilot was single or married. If he dove very close to the ground before he dropped his bombs, he was single. If he dropped his bombs from a higher altitude, he was married.

When did you go to Korea from Japan?
I got to Japan in August, 1950, and went to Korea in February, 1951.

What was Japan like?
It was wonderful. We had everything we wanted there. We had lots of time to go to town. When we were off duty, we could go at anytime. I was in the MP and we had martial law at the time, so we could arrest the Japanese. Later on, martial law ended and they could arrest us.

So, you were the police force in Japan. Did you have any trouble with the Japanese youths?
No, they were very humble. I went to Hiroshima before it was rebuilt. There were no adverse feelings when we were there. I think they felt that they should never have bombed Pearl Harbor.

So when you went to Korea, the good duty ended.
On the second day, they put me on perimeter guard duty on a machine gun and I didn't even know how to shoot it.

Was your base ever bombed?
Just by Bed Check Charlie, who would come over at the same time at night and drop mortar shells on us.

Was your friend, Mr. Wagner, in Korea at the same time?
We would write to each other and one time, he wrote, "I wish you would stop bombing our a**." There were times when our pilots would mistakenly bomb our own people. One time, I tried to find him. I knew what unit he was in and where

the unit was. It was a long way to go and when I went there, I couldn't find him and so I never saw him in Korea.

Let's talk about the trip you made back to [South] Korea.
A Korean professor told me about trips that were sponsored by the Korean government in appreciation for veterans of the Korean War. I organized a trip for five of us guys. U.S. Representative Marcy Kaptur, the veterans' friend, ironed out any problems we had. We had a chauffeur to take us any place we wanted to go. We went to museums, memorials and the DMZ*. While we were there, we went to Japan.

You said that Marcy Kaptur arranged all this? I've talked to a lot of veterans over the years and they all say good things about her.

Mr. Wagner—She's the best congressman we have ever had. She's a jewel. She helped a man I know who wanted to marry a woman in Shanghai. She cut through the red tape and took care of the whole thing.

Did you go again, after that trip?
Mr. Wagner and I went back again a few years later. We went to Camp Hialeah [Pusan*, South Korea] and then to a camp at Taegu. We had a hotel right near Seoul and walked into town every day.

What's at the Demilitarized Zone, the DMZ*, now?
We were right there where the armistice was signed. The heavily armed guards briefed us on the armistice. There is a team that goes into North Korea to find the Missing in Action, MIAs. It is still tense at the DMZ, after all these years.

When you got out, did you ever consider staying in?
No, they never me offered me any incentives like they offered other people.

What do you think about serving your country?
I'm happy that I served and proud that I served. You meet some people that you would never have met in civilian life. You have experiences that you would never have in civilian life. The experience cannot be duplicated.

U.S. ARMY

Gilbert Berry
U.S. Army

*At sixteen, Gilbert Berry
had lost both his mother and
father, which meant that
he had to quit school. He
worked at a job loading one-
hundred-pound bags into
boxcars, so basic training
and jump school were not too
tough for him.*

When did you go into the Army and where did you go for basic training?
I joined the Army in 1947 and was sent to Fort Ord, California.

How was basic training?
It was a little hard, but I thought the food at Fort Ord was fabulous, even though
most of the guys thought it was terrible.

Where did you go from Fort Ord?
I went to Japan with the 11th Airborne [Division, which included a parachute
infantry regiment] and took my jump training at Yamoto, where I got my wings.
Our unit was deactivated and returned to Fort Campbell, Kentucky, in 1949.

Why the Airborne?
When I told my brother that I was going to be a paratrooper, he laughed at me
and said that I would never make it. I never forgot that and I would have jumped
without a chute, if that's what it took to pass.

Where were you when the Korean War started?
When the [war] started, we were sent by ship back to Korea. We had spent thirty
days on the ocean to get home and now spent thirty days to go right back. We were
on a troop train going to California, which also had German prisoners, guarded by
black American soldiers. When we made a stop, the German prisoners went inside
to eat, while the black American soldiers had to stand outside and eat sandwiches.
I still can't get over that.

Where did you land?

We landed in Japan and from there to Kimpo Air Base in [South] Korea, near Seoul. We prepared for a jump into North Korea, forty-five miles north of Pyonyang, the capital of North Korea.

Let's talk about the jump.

We flew in C-119s [the Flying Boxcar air transport, built by Fairchild Aircraft] and our whole regiment of 3,000 men was involved. I was to be the BAR* man on that jump. We went out on patrols to locate the enemy and lost three men during that jump. We went back to Kimpo to train for another jump up north. I missed that jump, but drove the jeeps through enemy territory to the landing area.

What did you think about the North Korean soldiers?

I don't think they were well trained, but [they] were well indoctrinated. Life didn't mean anything to them. No matter how much we fired at them, they just kept coming.

What about the South Korean soldiers?

I didn't think much of them. You've heard of the term, bug out*. Well, every time we really needed them in a fight, they would bug out.

What did you think of your superiors?

They were fabulous. They were born to be leaders in combat. You would die for them.

What was the worst part of being in Korea?

It had to be the winter weather.

Did things change when the Chinese got into the war?

Oh yes. We saw the Marines when they were coming back from Chosin*. These poor Marines were in shock from being hit so hard and from losing so many. But they had all their dead with them in jeeps or on tanks. I'll never forget that about the Marines. . . . When we were advancing to the rear, we were supposed to burn everything, houses and buildings. There was a very old couple, whose house I was supposed to burn, but I couldn't do it. My sergeant gave me a direct order to burn their house, but I refused. I said that I'd shoot him before I burned their house. He said, "OK, let's go." I never would have forgiven myself if I had burned their house.

Did you have any R&R* while you were there?

I had five days in Japan. I ate at a big mess hall and got new clothes. But I forgot to bring some liquor back to my unit. Once on Thanksgiving, we had a frozen turkey left over from World War II.

How did you get back to the States?
My enlistment was up after three years, but I got what was called the Truman Year,* which meant because there was a war, I had to serve another year. I came back on another troop ship—another thirty-day trip and more standing in long chow lines. I was sent to Fort Custer [Michigan], where I was discharged.

What did you do when you got out?
I first went to work for my brothers in the sewer business. Then I went to work for Libbey-Owens-Ford for the next thirty-five years.

Did you use the GI Bill*?
No, but I used the VA [Veterans Administration] health benefits when I found out that I got frostbite in Korea. I now get disability every month from the VA.

Tell me about attending the Korean Church in Toledo.
Once a year, the Koreans invite the local veterans to thank them for liberating their country. After all these years, they are still grateful.

Mr. Berry has a chest full of medals for his military service and he is proud to have served his country, but he is sorry to see that, after more than fifty-six years, there is still no resolution to the war. He is sad for his friends who lost their lives in Korea and hopes that they did not die in vain.

Joel C. Davis
U.S. Army

Joel Davis served in the Army in Korea and helped to start an orphanage there for homeless kids wandering the streets of Seoul. That will be his military legacy.

Let's start in the beginning, before you were in the service.
I was born in East Toledo in 1929, graduated from Lake High School in 1950, and in April, 1951, I was drafted into the U.S. Army. Basic training was at Indiantown Gap Military Reservation in Pennsylvania. I got orders to ship to Korea, where I arrived in October, 1951. I spent the next seventeen months as an MP [military police] in Taegu.

Let's talk about an orphanage in Korea.
The streets and the alleys were full of orphan kids living in garbage cans. Our colonel was instrumental in starting this project. We not only provided a home for them, but also we fed and clothed them, and provided an education for them. When I left in February, 1953, the orphanage was self-supporting. We had built buildings and hired teachers and on Christmas, we had a Christmas party for them. This helped us to overcome our homesickness for not being home for the holidays. I am very proud to have been a part of that.

But there's more to the story.
In September, 1952, there was a wreck of a train bringing school children from Inchon to Seoul. The engine had blown up, killing seventy-five to eighty children. I was awakened and told about the wreck and responded by commandeering a put-put, which I took to the wreck. I could see the dead children lying alongside the tracks and the kids trapped inside the three cars. As we freed the children from the wreck, they were put on trucks, which I escorted to the hospital. I stayed to talk to the children in the hospital. And this happened on my birthday.

Do you have other stories to share about your service in Korea?
We accompanied the leader of South Korea, Syngman Rhee*, and General [James]
Van Fleet on a visit in our area. The South Koreans lined the railroad track to see
Syngman Rhee on the train and he, being very short, began jumping up and down
on the back of the observation car, waving to the people. General Van Fleet took
my rifle from me and said, "Soldier, hang on to the back of his pants, because if he
falls off this train, none of us will leave Korea alive."

There is a lot to talk about after you got out.
I am a member and a past commander of our local American Legion Post, a
member of the Veterans of Foreign Wars, and the Korean War Veterans Association.
The reason I want to continue working with veterans began when I was in Korea,
working at battalion level, with soldiers' problems. That has stayed with me since
leaving the service. I have also been elected to local, county, and state veterans'
organizations. I was honored to be made a member of the Ohio Veterans Hall of
Fame, the only one in Wood County.

What is the makeup of your local American Legion post?
It is mostly World War II veterans, but more Korean veterans are coming in. The
Vietnam veterans are very slow to join our organization.

Tell me about the Korean War Veterans Association.
I joined the national organization in 1986. I am number 276. In 1996, I transferred
my membership to the Toledo chapter and in the second year, I was the president.
Our treasurer got sick and died, so I took over that job, which I still have today.

What are the functions of your various organizations?
We turn out twelve to fifteen members for military funerals. We hold memorial
services at the funeral home or the cemetery.

I was at a KWVA meeting, where they talked about the schools they had visited.
We visit schools to tell the children about the Korean War. There is very little in
their history books, so they have a lot of questions. When they ask whether we won
the war, we answer, "Ask the South Koreans." Look at their modern capital city of
Seoul, which had been leveled in 1952. Look at their economy, the fifteenth largest
in the world [according to the 2010 listing by the International Monetary Fund].
And then ask if we won the war.

Korea has been called, "The Forgotten War," so it is our job to see that it is not
forgotten. More than 36,000 Americans died over there [NOTE: in 2000, the
Department of Defense stated the previous 50,000 casualty figure include all service
members who died during 1950-53. Of those, 36,516 were killed in Korea] and

we don't want them to be forgotten. I returned to Korea in 1996, sponsored by the Koreans. We paid for our transportation and they paid for everything else. We visit schools on a regular basis and are always welcomed by the children.

I see a plaque on the wall from the 40&8 Society.
The 40&8 Society* is the honor organization of the American Legion, which I joined thirty-eight years ago. We have a number of programs we fund, like the Nurses Training Program. I have helped a number of girls in Wood County become registered nurses. We also have a children's welfare program and a Flags For First Graders program. All are funded by the 40&8.

At this point, I read a letter from Ohio State Representative Randall Gardner, nominating Joel Davis for admission to the Ohio Veterans Hall of Fame, for his tireless work for veterans' causes, and for his work with Buckeyes Boys State, training programs, and the Ohio State Patrol Auxiliary. He was admitted in 2000. It is an honor to be elected to this organization, where only nineteen are admitted each year.

Herman R. DeBrosse
U.S. Army

Herman DeBrosse didn't have much trouble getting used to the Army routine. His mother died when he was fourteen and his father died when he was nineteen. His older brother had moved away then, so he didn't miss all the comforts of home that parents provide. He was used to being on his own.

I graduated from Macomber High School in 1947 and worked as a butcher until November, 1950, when I was drafted into the U.S. Army. Being in the Boy Scouts, sleeping outside, and the camaraderie helped me when I got in the Army. I went to Camp Breckenridge [Kentucky] for fourteen weeks of basic training. During our last week, I got the flu and went to the hospital. Out of our platoon, eight of us became a cadre. We got our first batch of recruits . . . black [soldiers]. Integration in the Army was rather new, but we got along fine and when they left after basic, they all shook my hand. So I went through basic training twice and when I was through, I had lost thirty-five pounds, from 220 to 185.

Where did you go from Breckenridge?
We went to Camp Stoneman, California, and shipped out from there to Japan.

How did you do on the ocean?
We ran into a storm with twenty-foot waves, but I didn't get seasick. After seven days, we reached Japan and shipped out immediately for Inchon*, Korea, on an LST*.

Where were you stationed in Korea?
I was assigned to the 24th Division, 19th Infantry Regiment, to a heavy 30-caliber machine gun crew. We were stationed on the 38th parallel*. Our gun was in a bunker and we lived behind it in the ground. In fact for the whole time I was in Korea, I lived in the ground. I had one shower in all the time I was in Korea.

How did you eat?
We ate C rations, but one Thanksgiving Day, we got a turkey dinner—the only hot meal I ever had in Korea. I had my mess kit full of turkey, potatoes, and yams and as I sat down, we had incoming. It was the first time I had been under fire, so

I hit the ground and so did the food. A sergeant came up and talked to me, but I couldn't hear and didn't hear until the next day. Most of the time, we ate cold C rations, but occasionally we would have time to heat the cans.

So you lived in the ground and ate cold food in cans all the time you were there.
We got used to it because we had no other choice. I decided not to fight it but learn to live with it.

What was the weather like?
It was cold, but much like the weather in Toledo.

Did you have adequate clothes?
We just had field jackets and regular Army boots. If our feet got wet, we would change socks, but we had canned heat [fuel made from denatured and jellied alcohol, burned directly in its can] in our bunker, so we could dry out. Koreans were sent up to our area to dig bunkers, one of them being a fourteen-year-old orphan boy who wanted to stay with us. So we adopted him and he lived with us in our bunker, which was strictly against the rules. He said that he shouldn't be there because he was a Christian and if captured, they would kill him. He taught me to sing "Silent Night" in Korean.

Mr. DeBrosse remembered and sang the first verse for the interview.

One night, I was in my sleeping bag when I accidentally spilled the canned heat on it. It instantly blazed up and I was trapped in the bag. Our Korean boy quickly leaped up and put out the flames and saved my life.

When did you leave Korea?
We were going through a creek bed, when we had incoming. My left knee was gashed to the bone. I was sent to a MASH unit*, where they sewed me up and sent me back to the line. Our unit was back off the line and was being sent back to Japan. I rejoined them and went to Japan.

Did you get to see Japan?
Only through the train window.

Where did you go?
Up north. I was assigned to troop information and education, to help GIs who didn't have a high school degree get their GED [General Education Diploma, a high-school equivalency degree]. We had to keep the guys busy, to keep them out of trouble in town. I also played on the division football team, the company basketball

team, the company softball team, and the division swim team. Another thing we did was to work at the orphanages. There were lots of orphans in Japan due to the war, and we did what we could to help.

How long were you in Japan?
I spent six months in Japan, or more time than I spent in Korea, and it was more fun.

You have now been away for about a year and are going home.
There is an interesting story about a man named Warren Detmer. We met on the train going down to basic at Breckenridge; he slept in the bunk next to me. When I was on the troop ship going to Japan, I met Warren again. He was in the bunk next to me. When I left Korea for Japan and got on the boat to return to the States, he was in the bunk next to me. When we got on the train to return to Breckenridge, he was in the same compartment. We left Breckenridge together and drove to Ohio together. We dropped him off in Napoleon and I continued on to Toledo. I never saw Warren again.

Have you joined any veteran's organizations?
No, I have never felt the need to rehash the war.

Any closing comments?
I'm glad I served, but I wouldn't want to do it again. I wish we could find another way instead of war.

Robert Fuller
U.S. Army

*In 1950, Robert Fuller and a friend wanted
to join the Navy, but they were only sixteen.
So they went to Frederick, Maryland, got
draft cards, and enlisted. His friend's brother
called the recruiter and told him that they
were sixteen. They were turned down and
returned home. In 1951, he got his draft
papers, which he tore up, and enlisted in
the Army.*

Where did you go to basic?
I was at Indiantown Gap Military Reservation in Pennsylvania, in the winter of
1951, and it was cold. I had eight weeks basic and eight weeks advanced training.
I got a sixteen-day pass and had to report to Camp Stoneman, California. We
shipped out from there to Japan and Korea.

What was it like on the troopship?
Everyone was seasick. A man in the latrine would be sitting on one stool with his
head in another one. Sixteen days later, we got to Camp Drake in Yokohama, Japan.
We stayed there about two weeks, while they were forming up units. I was assigned
to the Service Company, 2nd Division, 23rd Regiment.

Where did your unit go?
We were sent to Inchon*, where we had to wade ashore in waist-deep water. From
there, we were sent to relieve the 7th Division on Old Baldy*. The Chinese began
their big push, so the 7th couldn't get off of Old Baldy and we couldn't get up. This
was my initiation. I saw guys try to go up the hill and come down faster than they
went up.

Were you ever wounded?
No, but I had a scary experience. One night, I was hauling ammo to the front line,
when I ran the truck off the road. I hid in the field, when a Chinese patrol . . .
behind American lines, came by. I was sure they would find me and . . . my heart
was beating so hard I thought they could hear it. They didn't find me and the next
morning, I came out of the field, when [the Americans] came to rescue me. I found
out that I had been lying in a minefield.

Did you fight with South Koreans?
It was mostly Americans, but we had some [South] Koreans, called ROKs, and we also had a French division with us. It was a United Nations war effort. Some [groups] had Dutch and some had Thailanders and other UN troops.[General] MacArthur wanted to go across the border into China, in spite of President Truman's orders. The Chinese sent troops into Korea to stop the UN advance. As a result, Truman fired MacArthur.

Were you there when MacArthur was fired?
Yes, and I was there when President-elect [Dwight] Eisenhower came over. His son was up the road in the 3rd Division.

How long were you in Korea?
One year, four months, and two days. We had a point system. You got so many points for being on the frontlines, fewer for being behind the lines, etc. When we got sixteen points, we would be rotated back to the States.

What was the winter like?
It was cold. It sometimes got to forty degrees below zero. We had shoepacs*, but there was lots of frostbite and trench foot. We lost lots of people to the weather. I didn't think I had frostbite, but today, I have problems as a result of the cold over there.

Did the North Koreans adhere to the Geneva Conventions*?
No, but the Chinese did. Of course, if you were captured, you couldn't choose which one was to capture you.

Where did you live? In a bunker, a foxhole, or what?
We lived in tents with potbelly stoves. One time we got strafed and had to get in our foxholes. We found out it was American planes shooting at us.

Did the Koreans have any planes or tanks?
No, they just had rifles, machine guns, and mortars.

Did you have any leave time while you were in Korea?
I had one three-day pass to Seoul and seven days R&R* to Japan.

When did you come back to the States?
In October, we went to Inchon and sailed from there. I didn't get seasick, but we had a big storm in the Yellow Sea. It took us twenty days to get home and halfway home the engines stopped. A ship had sunk and we spent a couple of days looking for survivors. We didn't find any. We landed at San Francisco and flew from there to Camp Atterbury. Then they took us to Indianapolis to the bus station and our thirty-day leave began. When we got back, they sent us to Fort Campbell, Kentucky, where we were discharged.

How did this military experience affect you?
It was a good education, [one] that I would never get in school. I never got over the sight of little kids going through our garbage and the way the people lived in huts. There were no paved roads and no cars, just ox carts and bicycles.

Robert Fuller has had a distinguished career with veterans' organizations, including the VFW, the American Legion, Amvets, the County Veterans Service Board, and the Korean War Veterans Association. He has been a local commander and has held county and state offices. He has been active in talking to schoolchildren, telling them about the Korean War.

Joseph Heiny
U.S. Army

Joseph Heiny looked forward to being in the Army when he was drafted during the Korean War. Many of his older friends had been in World War II, so he tried to join up, but was turned down. He said that when he went in the Army, "I would do the best I could."

My family left Noblesville, Indiana, where I was born, to come to Toledo to find work. It was the start of the Depression and there were no jobs in Noblesville. My father got a job at E. W. Bliss [a machine and tool company], but was laid off and went to work for the WPA, Works Progress Administration, a government program begun by President [Franklin] Roosevelt. In 1937, my folks bought a house on WPA pay.

When I was drafted in 1950, I looked forward to being in the Army. I went to basic training in Camp Breckenridge, Kentucky, which had just been re-opened for the Korean War. The conditions there were such that the famous radio personality, Walter Winchell, would often close by saying, "Say a prayer for the boys at Camp Breckenridge."

Where did you go from Breckenridge?
We went to Cleveland, where we got a plane to Fort Lewis, Washington. After a week, we took a train to Vancouver, British Columbia. We got on a Canadian Airlines plane and flew to Japan. We landed in Tokyo, where we spent ten days getting orders. I was assigned to the 1st Cavalry Division. We went from Tokyo to Sasebo, through Hiroshima, which appeared to be about eighty percent cleaned up [after the atomic bomb drop].

Where did you land in Korea?
We landed in Pusan*, far south on the Korean peninsula. From there, I went to my unit just outside of Seoul, the capital city of South Korea.

What did Seoul look like?
It had changed hands a few times and had not been rebuilt. It looked like a bunch of tin shacks.

Where were the North Koreans, when you got there?

The 1st Cavalry was the first Army unit to go to Korea, since they were stationed in Japan. They took a lot of the punishment, when the North Koreans drove the UN forces all the way down to Pusan and almost into the sea. When I got there, they had been driven back over the 38th parallel*. I came after the UN counter-offensive at Pusan, the daring and successful UN landings at Inchon*, which cut off the North Koreans and chased them back north and the Chosin Reservoir*, when the Chinese entered the war [fall of 1950].

Where were you stationed?

We were near Uijeongbu, just north of Seoul, where we had an airstrip. I was stationed there for months, with HQ & HQ. I was a rifleman who had to become the company chief switchboard operator. I had six guys working for me . . . Signal Corps people, and there I am an infantryman. We got mortared and shelled numerous times and always in the evening. We had to bug out* three different times.

Bug out? Retreating?

Moving to a different place.

Were you always moving north?

We would move north and then we would turn around and come back south. We finally got to a place with an airstrip for small planes. The rest of my time in Korea was spent there. I made sergeant in one night when I was on duty. The commanding officer called to tell me that he wanted to talk to a battalion commander. I guessed correctly that he would want to talk to all his battalion commanders, so I had them waiting for his call. He got to talk to all his units, without the usual delay, and was so impressed that he asked who I was. I said, "Private Heiny, sir." He said, "Well today, you are a sergeant."

You have a Korean Service Ribbon with four stars. That means you were in four major battles. What were they?

I don't know. We never knew if the fight we were in was considered a major battle.

You also have a Combat Infantry Badge.* It is the highest honor that an infantryman can get. Although you were not in combat on a regular basis, you were under fire every day.

I am very proud of that Combat Infantry Badge.

Did the North Koreans have tanks?
I'm not sure if they were North Korean or Chinese [They were Russian tanks supplied by Joseph Stalin]. They were very small and not too useful, because of the terrain. They would usually be dug in and used as artillery.

How long were you at the front?
About ten months.

Were you advancing or holding ground?
Most of the time, we were just holding ground.

It appears that the plan was to just hold South Korea from the communist North, and if [General] MacArthur had continued into China, it could have caused a war with China.
I don't think that China was ready for war at that time, so they might have brought Russia into the war.

When did you leave Korea?
In December, 1951, we were relieved by a National Guard unit from Oklahoma. We went to Hokaido, Japan. We were there on Christmas Eve, so I went to a Catholic Church, where I heard the most beautiful hymns, sung in Japanese. I came home for thirty days and was sent to Fort Ord, California, until my time was up and I was discharged. I went back to work for DuPont [E. I. du Pont de Nemours and Co., American chemical company] until I retired.

You became active in servicemen's organizations.
On Christmas, 1999, my wife bought me a life membership to the VFW [Veterans of Foreign Wars]. Our post is active in community outreach. We go to the schools to tell the kids about the military and about the Korean War. I also do some community work with churches.

Mr. Heiny closed with this message to his family: Don't cry for me when I die. I enjoyed my life, so go to the post, and have a party.

John W. Hinds
U.S. Army

John Hinds and his twin brother joined the Ohio National Guard's 104th Field Artillery. They were specially trained in the Fire Direction Corps. They went from private to staff sergeant in less than two years and took what were called Series 10 Courses, in preparation to become officers. In December, 1949, they went to Columbus, Ohio, to accept their commissions as second lieutenants. They went to Fort Sill, Oklahoma, for two three-month courses in fire direction, sound direction, and survey. Their job was to locate the position of enemy artillery, from the flash as the weapon fired and/or the sound as it fired. Theirs was a very technical unit and very effective in finding unseen enemy gun positions.

You have mentioned the specialty courses you took, but have not said anything about basic training.

We had taken the Series 10 Courses and the schools at Fort Sill. When we were activated, we went into the regular Army as a whole unit. So there was no basic training and we all kept our National Guard ranks.

How long did you train in the States?

We trained from January to October and formed a 600-man unit to go to Korea. We landed at Inchon*, where we got clothes and equipment and then went to our unit position at Kumhwa [northeast of Seoul, now in North Korea].

How did you get your sound direction information?

Each morning we would crawl up to the top of the hills to get the coordinates, which we would send back to HQ.

Were you doing this in American-held territory?

Yes, we were behind the lines.

Did you ever see the enemy guns?

No, the Chinese were very careful to keep them concealed, often in caves. After we had been there for a while, they brought in a radar unit. It could pick up the enemy positions from their incoming shells. Of course, this was back in the 50s, so it was new to us.

What was the weather like?
Very cold, but we were issued the new Mickey Mouse boots [see shoepacs], which kept our feet from freezing.

Where did you live?
We were in a compound. We lived in a bunker and the enlisted men had a Quonset hut. We had a mess hall, but no showers. If we wanted a shower, we had to go back about five miles. There were always Korean kids around to wash our clothes and we had a fulltime medical tent with a person on duty at all times.

Had the main line of resistance been stabilized when you were there?
There was lots of artillery fire back and forth, but aside from some futile attempts at penetrating our lines, the MLR* was stable.

When did you go home?
In August we went to Pusan and were on the ship in the harbor, when the armistice was signed.

Did you know anything about the political situation at that time?
We knew a little. We didn't know why the Chinese came into the war and we didn't know what was going on in Washington.

Was [General] MacArthur still in command when you got there?
He had already been relieved of his command and was back in the States.

What did you think about your time in Korea?
It was sure different. I had left my wife and daughter back in the States, but I was stationed near my brother, who I could see about once a month. I also would occasionally see people I knew from back home. All of that helped.

You had been in the service for five years and were a first lieutenant. Did you consider re-enlisting?
I was told that I could go to a post in Okinawa and become a captain, but I wanted to get home to my family. If I had stayed in, I would probably have wound up in Vietnam.

Did you stay in the Ohio National Guard?
No, I got out of the military completely.

Earl R. Hufford
U.S. Army

When Earl Hufford went into the Army, he was given a series of tests that led to his assignment to Camp Pickett, Virginia, for eight weeks of basic training and eight weeks of medical corps training. From there, he went to Fort Sam Houston [Texas] for more medical training and then to Fort Leonard Wood, Missouri, for autopsy training. He then went to Fort Lewis, Washington, and on a ship to Korea. He had the top bunk of the five and was really seasick. His buddies brought him fruit and he was right near the head [the ship's toilet], which helped.

Where did you land in Korea?
We went into Inchon*, where I helped the corpsmen with the injured. After we got a foothold at Inchon, we went to Wonju [in north-central South Korea, ninety miles east of Seoul] where I worked in the Eleventh Evacuation Hospital. It was there that I ran the first dialysis machine.

What is an evac hospital? Is that where the injured went first?
They would first go to a MASH unit*, which was about two miles from the front and if necessary to an evac [evacuation] unit and then if they still needed more help, they were sent to the hospital in Tokyo. I felt for the wounded and their families, but I had to keep doing my job. In time, I became used to seeing injury and pain and although there was no time for sentiment, I always said a prayer for them.

Did you have a chaplain?
We had one chaplain to do the Catholic Mass and the Protestant service. I became the chaplain's assistant, helping with the services and driving him around. I also began a newspaper for the unit.

I see in one of your pictures that some buildings or tents were sandbagged. Were you subject to shelling or bombing?
Yes, they didn't care that it was a hospital. We finally had to take the red cross off the tents, because they used them for targets. We went to Inge, North Korea, to take care of the North Korean wounded. I guess we thought that all lives were important.

Do I understand that they were bombing your hospitals and you were building them a hospital for their wounded? Why didn't they build their own hospital?
Because they didn't care. They just left their men to die.

How long did you stay in North Korea?
We stayed a short time and then went
back to the evac hospital and then to a
MASH unit.

What did you do there?
I treated hemorrhagic fever, a disease
that causes internal bleeding . . . [it's]
caused by the mites that the rats carry
and we had lots of rats.

Did the enemy ever infiltrate your area?
I'm not sure if it was the enemy or not,
but someone stole my footlocker, with
all my clothes, the money I had saved for R&R*, and my rifle.

Did you go on R&R?
I went to Japan, which was just rebuilding. We bombed them and then we helped
them rebuild. What other country in the world would do that for their former
enemy and what country would build a hospital for their enemy?

How long were you in Korea?
I was there for fifteen months and it sure was wonderful to see the Golden Gate
Bridge [San Francisco], when we got back.

You were well trained in your field. Did the Army want you to stay in?
The Veterans Hospital in Ann Arbor, Michigan, wanted me, but I wanted to stay in
Toledo. I, and most veterans, eventually got two letters from the leaders of Korea,
thanking us for our service to their country. South Korea today is much better off
because of the help they got from America.

**I was at a Korean War Veterans Association meeting, where I heard a man
give a report on their activities with the local schools. Let's talk about your
membership and what they are doing for community outreach.**
I helped found the Toledo chapter and was the first secretary. I was a charter
member of the group, which now has 140 members. The history books in school
don't give much information about the Korean War, so we go to four or five schools
every year to talk about the war. It is not our intent to glorify war but to inform . . .
I dedicate this tape to all the veterans and the service men and women in the service
today. God bless them all.

Richard A. Janicki
U.S. Army

Richard Janicki had graduated from Libbey High School and was an engineering student at the University of Toledo. After he missed a semester at the university, he was drafted on February 7, 1951. He wasn't upset that he had to interrupt his education; he thought it was his duty to serve.

I went to basic training at Fort Leonard Wood, Missouri. Basic training was easy. We lived in a small barracks and had a small mess hall, so it wasn't like the rest of the camp with big barracks and mess halls that fed a thousand men. I was on my way to Aberdeen, Maryland, when I got pneumonia. I didn't remember anything until I woke up in a hospital in Baltimore. When I got out of the hospital, I went to Atlanta, Georgia, to the Atlanta General Depot. It was a school for track and vehicle mechanics. I was there less than six months, learning to work on all types of large equipment and construction equipment.

Did you go from there to Korea?
We boarded a troop ship in Oakland, California, and went to Hawaii. From Hawaii, we went to Japan and then to South Korea. We were on a big ship, so there wasn't much seasickness and the trip only took nine days. We got on a train that was full of bullet holes. It took three days to go 150 miles; then [we] got on trucks to go northeast of Seoul to Chuncheon and then north about forty miles. So it took three and a half days to go 200 miles.

What were your first impressions of Korea?
I remember that the hundred-year old trees were one inch in diameter. It was a zero growth country.

When did you get to Korea?
I got there in the fall [1951] and I spent two Christmases there.

Korea is famous for its cold weather.
We couldn't find enough stuff to burn to keep warm. I was assigned to a Light Equipment Combat Engineer Company. Our light equipment consisted of seven D9 Caterpillar bulldozers. A D9 dozer is twice as big as a D8 and a D8 could push ten houses at a time. They are tremendously big. We had shovels so big that one shovelful would fill up one and a half dump trucks.

What was this big equipment for?
We built an airstrip called K51. We took two mountains down and put them in a valley to build a 6,000 foot airstrip for jet planes.

Let's go back to the bullet-ridden train. Were you shot at on the train?
No, but we got a lot of mortar fire when we got to our destination.

As combat engineers, did you have to follow the infantry?
No, we had to go out ahead of them to build roads and bridges for them. While we were working, we got lots of mortar fire, but very few engineers were hit.

You were at the Chosin Reservoir* and in the combat zone near the Chinese border, weren't you?
Yes, not too far away. It was in enemy territory.

How long were you in the combat area?
I was there all the time that I was in Korea.

How do you deal with the constant fear when you are in a combat zone?
I had a lot of sleepless nights, but I had to keep busy to take my mind off of it.

Did they counterattack often?
At night it was very scary; there was mortar fire all around us. They tried to overrun us. They came close but never broke through.

Was there a regular tour of duty in Korea?
My time was running out, but I was still there. I was warned not to, but I went to report it to the adjutant general. And the next day, I left. On the trip back, we had a rough trip and everyone was getting sick. It took us fourteen days to get from Tokyo to Seattle. We spent a lot of time on deck gambling and I made a little money. We left Seattle and went to Fort Custer in Michigan. I had to go to a hospital in Michigan, because of the broken nose I got working on a truck.

When you got out, did you use the GI Bill* to buy a house or go to school?
I used it to go back to UT for the civil engineering course. They gave me so much money, which wasn't much, but a credit hour was only $14.00. I still have my GI insurance, but I never used the VA [Veterans Administration] for health care.

Laurence E. Kish
U.S. Army

When I was eighteen, I got my draft notice. Shortly after, a recruiting sergeant contacted me when I turned nineteen. He said that I could join up for the same two years as a draftee and I could go in right away and get it over with. So a friend and I decided to do it and off we went to Fort Jackson, South Carolina, for nineteen weeks. We then went to Seattle and boarded a ship for the Far East Command. We were on the ocean for two weeks and I was sick the whole time. I missed seventeen straight meals.

We went to Camp Drake in Yokohama, Japan, to pick up our gear and crossed the sea to Inchon*, South Korea. We had to anchor five miles off shore and take landing craft in to shore. When we got on a train that night, a sergeant asked me if I knew how to operate a machine gun. I said that I did, so he put me at the end of the car with a machine gun and there was another man at the other end of the car with a machine gun. It seems there were lots of guerillas between Inchon and Chuncheon, where we were going.

I was assigned to the 40th Division, 225th Infantry Regiment. We were to replace the California National Guard, which was going home. In September, 1952, we were sent to the main line of resistance at the Punch Bowl*, which was an area surrounded by mountains. As we came up to our position at night, we saw tracers flying everywhere. I was asked to take apart a machine gun and after I did, I became the weapons squad leader. Our first squad members were five Koreans and seven Americans. I remember one night the enemy was playing American hit songs. They were playing "Vaya Con Dios," [Go with God] with Les Paul and Mary Ford. The next night, the bugles started blowing, the cymbals were clanging, and we were in for a ten-hour attack. This was our first attack, after just three days on the line.

When did this take place?
It was in October and the cold winds were blowing in from Manchuria. It was about forty below zero. We had adequate clothes, but only regular leather Army boots. The medics came every day, to make sure we changed our socks every day.

What kind of soldiers were the Koreans?
They were good soldiers who worked well with the GIs. The North Koreans were not as well trained as the Chinese. They had Russian burp guns, which held 160

rounds, while we had M1s [semi-automatic, .30-caliber rifle], which held eight rounds. We wanted to get hold of M2 carbines [a fully automatic rifle], which had a 30-round clip.

How did you feel when you heard the bugles blowing?
We were frightened to death, but adrenaline takes over. We were always scared; fear never leaves you in combat. They were always trying to infiltrate, but we put out wire with beer cans filled with stones, which would make noise. We also had hush flares, which were shell casings filled with napalm. When they were tripped, they shot napalm into the air.

Did you have air support?
Marine pilots would fly Corsairs [Douglas A-1 Skyraider] in from carriers to drop napalm. The Australians used our P51s* for strafing and dog fights. The Chinese had Russian MiG* jets, which were the best fighters in Korea. We had the F84* jets, which came from the airfields further south.

Let's talk about your second major battle.
It was in early December, when a large force of Chinese wanted to take over our strategic position, thirty miles north of the 38th parallel*. I took a squad out on a patrol. We were out there for fourteen hours. It was about fifteen degrees below zero and my hands and feet were freezing. Some of our guys got frostbite that night. Our CO wanted us to shave every day, so we used beer and Noxzema.

Did you have any heat in your bunkers?
We were supposed to get four gallons of fuel oil every night, but the rear echelon took most of it. We were lucky to get one gallon per night. We would burn wooden ammunition cases.

How did you get the [M2] carbines?
One of our guys went back for a shower, where he found that the people behind the lines all had carbines. When he came back with a carbine, I asked him how he got it. He said that when people went into the shower, they left their weapons outside. So when he went into the showers, he left his M1 and when he came out he took a carbine. We sent him back to the showers, until we all had carbines.

Did you win this second major battle?
Yes, we won all three of our major battles.

Were your tanks useful in Korea?
They were used only for support and backup. A sergeant took me to the MLR*, to

look down into the valley with binoculars, where I saw all these black specks there. They were tanks that got bogged down in the rice paddies, during the Chinese offensive. There must have been sixty or seventy that were destroyed.

I finally got promoted to private first class; two weeks later I got promoted to corporal, one month later to sergeant, and three months later to sergeant first class. I had only been in Korea from August to February and already, I was an SFC. What had happened was that one sergeant went home, three were wounded, and three were killed, so they needed a platoon sergeant. In July, 1953, when I was rotated home, I got another promotion to master sergeant.

I met a Catholic chaplain, who would come up to the MLR for Mass. About twenty of us were going back to Mass, when we saw about fifty Koreans who were going to Mass. I didn't know it, but there are a lot of Catholic Koreans. This chaplain had earned a Silver Star, when he and another fellow went out under fire, in the front of the line, to rescue some wounded GIs. I eventually became a Mass server. He knew who the Catholics were, so if they didn't come to Mass, he went looking for them.

So religion was very important in war.
Yes, he said if you weren't a Catholic, come anyway. We are all children of God.

How was the food?
We had one hot meal every day and the rest were C rations. On Christmas, we had a hot meal, even though it was thirty degrees below zero. I remember that my coffee froze while I was eating.

Did you get regular packages and mail?
Yes, we got regular mail. My mother knew where I was. I don't think they were censoring the mail. I lost three good friends in Korea. One was a medic who sat down on a land mine. Another was killed in a gun fight while on patrol. Another was killed on Memorial Day. There was another friend who was getting ready to go home because his wife was dying. As he was going into his bunker, he was hit with a mortar and died.

I was going on patrol one day when the CO said not to go because I was going home. I went back below and got new clothes and went to Inchon. When the boat came in, out came four friends from Toledo. They had waited to be drafted, instead of joining, as I had. They were just getting into Korea as I was leaving and they had to stay two years. On July 27, 1953, we left Korea, the day the armistice was signed, and went to San Francisco. On August 12, I arrived in Toledo, and on August 22, I was married. I had six months to go, which I served in Fort Jackson, South Carolina.

After I got out, I started going to reunions to see the people I had fought together with in Korea. I also joined the National Korean War Veterans Association and the Northwest Ohio Korean War Veterans Association.

Robert Lempke
U.S. Army

Robert Lempke, who was orphaned at a very early age, joined the National Guard at fifteen, with the consent of his grandparents. When his unit was activated, he was sent to Camp Polk, Louisiana. In November, 1950, he went to Camp Stoneman, California, and by ship to Japan. After five days in Japan, he was sent to a replacement depot in South Korea, assigned to the 2nd Division, 23rd Regiment.

We were in numerous battles at Old Baldy*, Pork Chop Hill*, Five Fingers, and Outpost Harry*. We were on line for about eight months and taken off for thirty days and then [sent] right back to the front, where we were attached to a French battalion. We were in probably eight battles back and forth from South Korea to North Korea and back. I lost a lot of friends in that time. In one action, we lost all but four in our platoon. We were captured by the North Koreans for about ten days, until we were freed by the Marines.

Were you at the Chosin Reservoir*?
No, but we were not far away. Our losses were on Old Baldy and Pork Chop Hill. I remember the first day, when seven of us replacements were taken up to the line. We heard artillery fire and all kinds of noises and we were scared. I also will never forget how the North Koreans and the Chinese would just keep coming, no matter how many we killed. Outpost Harry was the bloodiest battle I was in. There must have been four or five thousand dead out in front of us. We had a shortage of officers, so the battalion commander wanted me to take a lieutenant's position, but he took it back when he found out that I wasn't eighteen.

Where were you when the fighting stopped?
We were on the front line. The war was supposed to stop at 7:30, and at 6:30 we got one of the biggest barrages I ever experienced.

Where were you discharged?
I went to Fort Lewis, Washington, and Fort Sheridan, Illinois, for ten days and then home to Toledo.

Surely you were given an opportunity to re-up.
No chance. My draft board couldn't believe that at my age, I had been to Korea and back.

Have you joined any of the veterans' organizations?
Yes, I was very active in the American Legion. I am a past commander and I worked on the widows and orphans welfare foundations for the state of Ohio and on American Legion baseball. I went to a few of the 2nd Infantry Division reunions, but it is hard to see those who lost limbs or their sight.

What was your civilian career?
I worked for forty years in the insurance industry, retired, worked in the security business for five years, and then back to insurance.

Terence J. Mohler
U.S. Army

Terence Mohler looked upon his military service as a duty and an honor, as had his father in World War I and his brother in World War II. "Now it was my turn to serve," he said. He was drafted in March, 1951. Thinking that he might fail the sight and hearing test, he memorized the eye chart and since there was no hearing test, he was inducted into the Army. He took his basic training at Fort Bragg, North Carolina, and trained to be a radio operator. After basic, he spent a week on a train, going to California and boarded a troop ship for the Far East Command.

Describe the trip on the Pacific Ocean.
We were in a storm for a week, so we couldn't go on deck. As a result of seasickness, we were all on a diet. We went to Yokohama and then to Pusan*, South Korea. The trip to Pusan on an LST* was worse than the storm in the Pacific. It took us three days to go 400 miles in rough seas. Pusan is [as far south as] the North Koreans had driven the UN forces, before they were forced back across the 38th parallel*. There was nothing left of the city.

To what unit were you assigned?
I was in the 623rd Artillery Battalion, Battery C. We had six 155-mm howitzers, which we took up to the MLR*. We lived in bunkers in the ground. We covered the top with trees, which we cut down, and rocks and dirt over that.

What was the situation at the front?
The 623rd was not adequately prepared. We were outnumbered and outshelled. Sometimes, we only had three shells per artillery piece. We had old shoepacs* from World War II and our clothing was not adequate for the Korean weather. Our sleeping bags were OK, so the men in the foxholes would be sitting in their sleeping bags to keep warm. We were often subjected to sniper fire and when we shelled them, they would return three times as much fire.

I had trained as a radio operator, but we didn't have any radios. We communicated by switchboard and telephone. One of my jobs was to take care of the mail, which was very important to the men. I would not only deliver the mail to them, but also read it to those who were illiterate. I was also in charge of our supplies of tobacco, beer, and soap. The beer became a problem, because the non-drinkers would sell their ration to the drinkers and that sometimes created problems of intoxication. Sometimes, I felt that I was fighting two wars, one with the North Koreans and one over the beer supply. We ate mostly cold rations and often under fire. We wore the same clothes from December to March and we used our helmets to hold the cold water we got out of the river for shaving.

One day, I was driving alone in a jeep when I hit a rock or a land mine. The jeep went one way and I went another way. As a result, I left on a stretcher. I went by helicopter, plane, and ambulance from Korea to Japan to Guam to California and finally to Percy Jones Hospital in Battle Creek, Michigan. It took three surgeries before I could walk with crutches and two years before I could walk without crutches.

I would like to ask you some more questions about Korea. You said that you lived in bunkers; how did you heat them?
We used diesel fuel in our stoves. It didn't take much to heat them with seven or eight feet of trees, dirt, and stones for a roof.

You mentioned illiteracy. I didn't think the Army would accept an illiterate.
They were National Guardsmen whose units had been activated, so they weren't tested.

We had earlier talked about the Korean War as the forgotten war. Your comment?
It was a mistake to call it a police action and it didn't get the media attention it should have. We had also talked about the possibility that the people had used up their patriotism in World War II, which been over for just five years.

And when you returned to civilian life, did you make use of the GI Bill of Rights*?
Yes, I returned to college, for which the GI Bill paid.

Let's talk about your work with the Burial Corps.
At the request of the families, we provide a military ceremony at the gravesite, after the service. We fire a salute to the deceased, play taps, fold the American flag, and give it to the spouse. I belong to the VFW [Veterans of Foreign Wars], but the Burial Corps is a separate organization from the VFW or the American Legion. As a VFW member, I am one of those who visit the schools to talk to the students about America's wars and about the military. I am the person who talks about the Korean War.

What are the students' reactions?

They know very little about America's wars, since the textbooks exclude or give very little information about the subject. They are very interested in hearing from us. [Americans should] join the Peace Corps or the military service, one of the two, to establish a more realistic view of what the world's all about.

Donald Proudfoot
U.S. Army

Donald Proudfoot said that he was just a country boy who had never been far from home, so the Army was quite a change for him. Since he had been drafted, he wasn't too enthusiastic to begin with. He said that the training was pretty intensive, because they were trying to get them ready for Korea as soon as possible.

Where was basic training?
I went to Camp Breckenridge, Kentucky, for basic and then for reasons known only to the Army, I was sent to cooks' school for eight more weeks.

Where did you go next?
We went to Fort Lawton, Washington, where we got on a ship for Japan. The second day out, we got into a big storm and all 1,800 troops in the bottom of the ship were sick. There was heaving and dry heaving and we couldn't eat for a couple of days. We finally got to Yokohama, Japan, where we stayed for a week on an old Japanese army base. After a week, we sailed for Inchon*, Korea.

What were your first impressions of Korea?
Not very good. We got on an old-fashioned train that had wooden seats and all the windows shot out. After a week in a replacement depot, I was assigned to a tank battalion, just behind the front lines.

Were the Chinese in the war at that time?
Yes.

Were you under fire?
We were under fire for almost the entire time I was in Korea. We were also behind a unit of Korean soldiers, which didn't make us feel too secure, because they might get up and leave at any time.

What was your job in the tank battalion?
I was a cook.

Did you cook for the units on the front lines?
The tankers on the line got hot meals most of the time. Just a few times did they have C rations.

Did the enemy have an air force?
When I got there, their air force was about done. We got strafed just once by a lone North Korean plane. Our air force dominated the air and we didn't see any Russian MiGs*.

How did your tanks operate in the mountains of Korea?
Our tanks used to go out into no man's land, but on July 4, 1952, a unit of tanks got trapped out there and we lost a number of men and tanks to the Chinese. From then on, the tanks would dig into the hills. They had 105mm guns, so they would be used as artillery.

Did the North Koreans and Chinese have tanks?
No, but they had artillery, lots of mortars, and a lot of people. Before I left in February, they needed more people on the line and I was one of those who went up. The ground was so frozen that we couldn't dig in. We had good sleeping bags and good shoepacs*, which were better than regular boots, but we just wore fatigues, we didn't have long johns. And that far north, it could get to forty degrees below zero.

What was it like inside a tank?
In the summer it was like an oven. In the winter it was very cold until the engine warmed it up inside.

Did you feel that the UN forces had the situation under control?
Yes, we were mostly in a standstill war. We just moved a few times in the year that I was there.

What did the GIs think of their commander, General [Douglas] MacArthur?
We thought he was flamboyant with his pipe and sunglasses.

While you were there you had R&R*. Where did you go?
The combat units got a week and I chose Tokyo, because I knew an American family who lived there. I was able to go to the PX [retail store on a military base] and buy Christmas presents to send home. The family showed me around the city, which at that time looked pretty backward, although the shopping areas looked modern. The Japanese people were very friendly and sometimes, the GIs took advantage of them, but they did like the GI's money.

What was Seoul like back then?
It was a mess. It looked like the war zone that it was. Today, of course, Seoul is modern, with skyscrapers and freeways.

When your year was up how did you go home?

We went by boat to Sasebo, Japan, where we stayed for two weeks, then we took a ship to Seattle, Washington, and a train with Pullman cars to Camp Custer, Michigan. We didn't have any bad weather like we had going over, but we were still down in the bottom of the ship sleeping in bunks, five high. For Korean veterans, there were no bands playing. We just went home and took off our uniforms.

What do you think about serving your country?

It helped me and I think that if young men and women had six months in the service, it would help them. It takes the attention off of you and makes you a better citizen. . . . I can see why our country is trying to keep peace in the world. Somebody has to take the lead and our country has done it. The world is a better place because we have taken the lead.

Albert W. Reiser, Jr.
U.S. Army

While in basic training, Alfred Reiser volunteered to be a cook to get out of the weather. After burning up twenty-nine pork chops, it was mutually agreed that he was probably not cut out to be a cook and that he should go back to his unit.

In November, 1960, at age twenty-five, I was sworn into the U.S. Army. I was sent to Camp Breckenridge, Kentucky, where we drilled in the rain and snow and where many recruits got pneumonia. From there, I went to Fort Belvoir, Virginia, for an eight-week engineer school. I was picked out of 400 to go to Leadership School there, which was very difficult. We started with thirty-five, but only nineteen passed. Twelve signed up for Officer Candidate School. I didn't because it required that I extend for two more years.

When you turned down OCS, where were you sent?
I got orders for Seattle, Washington. And from there to Japan, where I joined the 5th Regimental Combat Team, which consisted of three companies. I got up to the front in July, 1951. I reported to the 1st Battalion, where I was put in communications. My job was to lay wire from the outposts out in front of our lines, back to our main line of resistance. Wires were constantly being destroyed by enemy mortars. I was on the switchboard for a while and then they needed a radio operator. I said that I couldn't do it because it takes an eight-week course to learn to operate the radio. They said that I had to learn in a hurry, because they didn't have a radio operator. So I did learn in a hurry and was assigned to the battalion communications officer, to operate his radio and drive his jeep.

Were you there in the winter?
Yes, we were supposed to get the new boots [the so-called Mickey Mouse boots. See "shoepacs" for more information] but they never came. It was said that the people back behind the lines got the boots and the guys on line were getting frozen feet. I might mention that that winter, the Chinese would play Christmas songs to try to make us homesick.

Let's back up so I can ask you some questions. How did you get to Korea, ship or plane?
We went by ship up toward Alaska and took the northern route to Japan. We were on an LST* from Japan to Korea when we got into a terrible storm. It was so bad that the chains that held our bunks broke and one of the LSTs had its seams buckled.

You were in Korea for nine months, so you saw all the seasons, didn't you?
I was there for Thanksgiving and Christmas. They did bring us up some turkey on Thanksgiving, which I ate sitting in the muck.

Some have said that it got to forty degrees below zero. Were you dressed properly?
I'd say that aside from the boots, we were dressed OK.

During your time in Korea, did you go on R&R*?
I went to Japan for R&R. The Japanese were very friendly to the GIs. I had accumulated a lot of money, so I had a great time in Japan. I was the driver and radioman for the battalion colonel, so we would often go way past the 38th parallel* and I was privy to some secret information. I had to be very careful, when my buddies would ask me questions, that I didn't divulge any privileged information. The Army had done a background check on me all the way back to my father. I had to be cleared because certain information came to me like codes, when we might be starting a push, and reports from the patrols that returned at night.

When you came home, was your time up?
No, I was sent to Fort Custer [Michigan], where I worked in the message center. I was released after twenty-one months, but I had a five-year reserve commitment.

Let's talk about your activity with the Korean War Veterans Association.
We funded a Korean War monument which [is] in a prominent square in downtown Toledo, Ohio. The KWVA visits the schools every year to inform the children about the Korean War. I am also vice commander of a post in Sylvania, Ohio.

Do you have some closing thoughts for your family?
All I can say is that God was with me.

Donald W. Swain
U.S. Army

Donald Swain said that basic training was rough, but having grown up during the Depression, he was used to rough times. He lived on a farm in Macksburg, Ohio, with twelve brothers and sisters. They hunted and fished and lived off the land. After high school he had lived away from home, so going to basic training wasn't too bad.

After basic training at Fort Knox, Kentucky, I went to Fort Lewis, Washington, [then] on a ship to Japan. We took the northern route up near the Aleutian Islands where we ran into some rough weather. We landed at Kobe, Japan, where I was picked to go to cooks' school. I guess they picked me because I had worked for an ice cream company. I spent two months in cooks' school and then in May, I went to Kimpo, South Korea, where I never did cook, but became a rifleman. In Kimpo, I was assigned to the 1st Cavalry Division, a distinguished fighting unit. For two months, we guarded a camp and then went to the front.

Were you there at Chosin*, when the Chinese entered the war?
I was south of Chosin, but we did engage the Chinese later.

Were you under constant attack by the Chinese?
They would attack us in the early morning or late at night, and in between they would shell us. They didn't have many planes, but every night, a small plane would come over and drop a small bomb. We called him "Bed Check Charlie."

Where did you live?
We lived mostly in trenches and foxholes.

How were the bugs?
Mostly lice—we all had them.

What was the weather like?
It was OK in summer, but the winters were really cold. We had parkas and shoepacs*, but there was still lots of frostbite.

When were you wounded?
I was hit by a sniper who was hidden under an evergreen tree. I went back to the aid station and then to a Swedish Red Cross hospital. I stayed there for six days and after thirteen days I was back with my unit.

Were you constantly advancing?
The going was pretty slow, particularly when we were going up Old Baldy*.

Did the enemy have artillery and tanks?
They had small tanks and lots of artillery.

Did the Americans dominate the air?
Yes, we got lots of help from the Air Force. They dropped bombs and napalm on the enemy lines.

Did you ever get up to the Yalu River*?
No. I was on KP [kitchen duty] when an officer told me that I was leaving the next day. So I packed up and was sent to Japan. I thought that I would be home by Christmas Eve. We left Camp Stoneman [California] and had the wonderful experience of spending Christmas Eve in the Chicago train station. I got home at 2:30 p.m., on New Year's Eve.

When you were being discharged, weren't you asked to re-up?
No, no one even asked me.

The GI Bill of Rights* allowed you to buy a house with a low down payment and have a free education. Did you take advantage of that?
No, but I still have the GI insurance that I got when I went in the Army.

Do you use the VA [Veterans Administration] health benefits?
Yes, I have used it for many years.

Do you belong to any veterans' organizations?
I am a life member of the VFW [Veterans of Foreign Wars] in Bellevue, Ohio, and the American Legion in Monroeville.

Robert L. Swain
U.S. Army

Robert Swain lived in Elba, Ohio. At sixteen, he was in high school when his parents gave their consent for him to join the Army. He went for basic training to Fort Knox, Kentucky, which he didn't think was too hard, but that's because he was sixteen. After basic, he went home for thirteen days and, in spite of being told by the recruiter that he would not be sent overseas, he went to Japan.

We were on the ocean about two weeks and although I had never been on the ocean before, I didn't get seasick. We landed at Tokyo and were stationed nearby at Camp King.

Did you get into Tokyo?
We could get to Tokyo in about an hour by train.

How were you treated by the Japanese people?
We never had a problem with them.

How long were you in Japan?
I had been there about a year and was due to go back to the States, when the Korean War broke out. The 24th Infantry Division, 8th Army, was the first unit sent to Korea from Japan. I was with the 1st Cavalry Division, 8th Regiment, which followed the 24th into Korea. We took the short trip from Japan to Korea. When we got off the boat, we began to walk about five miles through small towns to get to the front. I was eighteen and I was now in combat.

Were you all new to combat?
Some of the sergeants and the officers were veterans, but most of us were green.

When did you go into combat?
It was late June or July, 1950.

Was this a battle of tanks and artillery?
Where we were, it was soldier against soldier, with just small arms fire.

Had the Chinese entered the war?
No, I was gone before they entered the war [October, 1950].

When were you wounded?
It was in August. We were in a rear rest area when a guerilla force hit us just before daylight. I was wounded and taken to an aid station. I heard later that most of our platoon was killed.

So this was not the usual kind of war?
No, they would dress like civilians and come through our lines.

Where were you taken after the aid station?
I was evacuated by train to a field hospital and then was flown to the American Hospital in Tokyo. I stayed there a few weeks and was flown to Washington, D.C., to Walter Reed Hospital. I spent eight or nine months at Walter Reed. I had lost my right thumb and had a chest wound.

Were you discharged after that?
I went before a board with a Veterans Administration representative to establish my disability and the compensation I was to receive.

This all had to be a very scary experience for a person who was still a teenager.
My wounds were not that serious compared to some that I saw at Walter Reed. I was lucky that I still had all my limbs and I could walk.

Do you have any regrets about being in the service?
No, it helped me grow up in a hurry. I got out in May, 1951, and got a job on the railroad, where I spent the next forty-three years.

Any last words before we close?
No, I think we've said it all.

Thomas J. Van Buren
U.S. Army and U.S. Air Force

Thomas Van Buren began the Air Force tradition in his family. He spent twenty years in the Ohio Air National Guard, after serving in Korea with the U.S. Army. His son became a career officer in the U.S. Air Force as a missile officer and his grandson is a crew chief on an Air Force gunship.

In September, 1951, I was drafted into the U.S. Army and was sent to Fort Custer, Michigan, and on to Fort Leonard Wood, Missouri, for basic training and advanced engineer training. Basic training was not too bad, because I had been a Boy Scout and had been away from home before. But the weather at Fort Leonard Wood was so bad that it was nicknamed "Little Korea." After basic training, I was sent to Fort Bliss, Texas, to antiaircraft artillery school. I had been an apprentice to a gunsmith back home, so with my background, I trained to be a gunner and an artillery mechanic. There was only one mechanic per battery, because it was a highly skilled job.

Back in 1951 and 1952, there were two places where you might go. One was EUCOM, the European Command, which usually meant Germany, or FECOM, the Far East Command, which usually meant Korea. Where did you go?
We took a troop ship to Camp Drake, Japan, where we stayed a few days and then to Yokohama, where we took a two-day trip, on the Sea of Japan, to Inchon*, South Korea. I was assigned to Battery B, 15th Antiaircraft Artillery, 7th Infantry Division.

Did you go right up on line?
We were issued flak jackets and steel helmets, spent a few days behind the lines, and went up on line, where I stayed in a bunker for a week.

Were you under fire all that time?
We were bombarded daily by Chinese 122mm mortars.

You were bombarded by the Chinese?
The Chinese had entered the war in late '50 [October, 1950]. The 7th Infantry Division had been the rear guard at Chosin Reservoir*, when the Chinese came over the border and overwhelmed the UN forces. That was before I joined the unit.

How did you feel when you had your first taste of combat?
Scared as hell. They attacked mostly at night and would use loud horns and whistles.

Were you in constant combat?
We would pull off the line for two or three days and then go right back.

This was a UN action. Did you see troops from other nations?
There were seventeen UN countries in Korea, [including] Ethiopians, Colombians, Dutch, New Zealanders, and the English Commonwealth countries, almost all of whom were infantrymen. The U.S. supplied the heavy weapons and in fact, most all of the weapons.

Tell me how you got wounded.
It was in August or September, 1952. We were in a rear assembly area, cleaning up, when we got incoming. I was in a tent taking a bath when it started. I was naked, but I was running across an open area, with my new, clean clothes in my hand, when I got hit in the left thigh. A medic came and cut open my new fatigues to get to the wound. I spent a few days off duty, then went back to the front line for three months.

How did you eat and sleep on line?
We ate mostly C rations, but we got a few hot meals. We slept in bunkers on stretchers suspended by communication wire. It was warmer than sleeping on the ground. We wore layers of clothes and parkas. When it was real cold, we wore what we called Mickey Mouse insulated boots* [see "shoepacs"]. There was a lot of frostbite on fingers, toes, ears, and noses. If we were in defensive positions, we slept in foxholes, where the nights were long and scary.

Did you get any R&R*?
I went to Japan for six or seven days. We were treated very well by the Japanese. I don't think I was ripped off by anyone in Japan.

How long were you in Korea?
Twelve or thirteen months. I was discharged at Fort Sheridan, Illinois. When I got home, I still had an eight-year obligation in the reserves. I found out that the University of Toledo had the 199th AAA [Antiaircraft Artillery Automatic Weapons Battalion], the same kind of unit in which I had been in Korea, so, I joined the unit. I was the small arms instructor from 1953 to 1959. I had been a corporal when I left Korea and eventually became a master sergeant at UT.

That wasn't the end of your military career, was it?

No, after a twelve-year break, while I worked three jobs, I talked to an Air Force recruiter, who told me there was a need for a person with my weapons skills. I joined the Ohio National Guard. I joined as a staff sergeant and in two months, I was promoted to tech sergeant. I spent twenty years at Toledo Express Airport, with the 180th Air National Guard, teaching all the small arms classes and retiring as a master sergeant.

Richard S. Wagner
U.S. Army

On April 18, 2006, Richard Wagner and his longtime friend from school, Richard Piriczky (his comments are in italic), were interviewed. Each helped the other to remember what took place fifty-four years ago when they were both in Korea. Mr. Wagner tried to enlist in the Army at age fifteen, but was turned down. He did join at sixteen, with the help of a neighbor lady, who signed for him when his mother wouldn't. During basic training, they found out that he wasn't the minimum age of eighteen, but instead of putting him out, they let him sign a waiver to stay in.

Where did you go to basic training?
I went to Fort Knox, Kentucky, and for advanced training, I went to Fort Benning, Georgia. On June 25, 1950, the war broke out, so I was sent to the West Coast and on a ship to Sasebo, Japan, where I stayed for a month.

Where did you go in Korea?
We landed at Pusan*. I was assigned as a rifleman, to the 24th Division, 21st Regiment, and got into the fighting with the North Koreans at Taejon* in South Korea.

I understand that you were put up for a Silver Star.
There were three squads trying to take a hill, but we were pinned down by a machine gun nest. The first sergeant ordered me and another man to take out the machine gun, which we did with grenades and automatic fire. I was told that I was put up for the medal by the sergeant and our officer, but apparently in the chaos, it didn't happen. Later on, the first sergeant couldn't verify it because he passed away, and my records burned in the fire in St. Louis.

Let's talk about your Purple Heart. How did that happen?
In 1951, I was in a squad that was taking a hill, when we were hit by an artillery shell. All my buddies who I had come over with were killed but me. Everyone was killed but me. I was sent back to a field hospital for a week and then back on the line.

You have five battle stars. Where did you earn those?
I can only remember one at the Pusan perimeter and one crossing the Nakdong River [South Korea]. We were supposed to cross in the dark, but didn't start until eight a.m. The river was only one hundred yards wide, but I saw more dead soldiers there than I have ever seen. I don't remember the rest, because we were not always sure whether the fight we were in was considered a battle. The most important award I earned was the Combat Infantry Badge*. That meant that I was a frontline soldier. I remember being in a foxhole, with a buddy, under fire for about fifteen hours, and thinking that this war made him closer to me than my own brother.

How was the winter in Korea?
We were up near the Yalu River* in October. Our clothes and our boots were not sufficient. We had those old shoepacs* with the felt liners. Our feet would sweat during the day and then would freeze at night. There was a lot of frostbite. There were cases of people losing their feet or their fingers or their ears. We also had the problem of old ammo from World War II. We had lots of duds.

Were you in the high country?
I wasn't at Chosin*. I was west of there, but wherever you were, the weather was the same.

It was at Chosin that the Army and the Marines were said to be making a strategic withdrawal.
Someone said that they were "fighting in a different direction." I thought that was B.S.

Did you fight the Chinese?
Yes, they came into the war before MacArthur knew it. It was about September or October, 1950 [October].

When all this was happening, you were still a teenager.
I was eighteen in October, 1950.

You had all those medals and you still weren't old enough to vote or to drink alcohol. You sure saw a lot of life before you were eighteen. I understand your friend Richard Piriczky, who was in Korea at the same time, came looking for you.
My CO asked me how long I had been in Korea. I didn't know, so we looked in my records and found out that I had been there fourteen months. I was supposed to go back to the States after twelve months. Also, when we opened my file, there was a note from Richard, telling me he had been there and couldn't find me.

Did you leave Korea then?

I was sent to Japan, where I spent thirty days, before I was shipped to Fort Custer, Michigan.

You had frostbite. Was it a permanent disability?

Yes, I qualified for disability from the government.

Back to Korea. You were the youngest guy in your outfit. Did the older guys look after you?

No, age wasn't a problem for me.

Mr. Piriczky—You were expected to carry your load regardless of your age.

You were not yet twenty when you were discharged. Did you think about staying in?

No, I had seen enough of war.

You said earlier that you went back to Korea.

Mr. Piriczky—We went back in 1998. Congresswoman Marcy Kaptur helped us with the arrangements. We saw Korea and also Japan, where we got VIP treatment. We went to Hiroshima, where we saw Peace Park and rang the peace bell.

Mr. Wagner and Mr. Piriczky have been friends since their childhood in Birmingham, East Toledo. They survived the Great Depression and the Korean War and are friends to this day. Both have the honor of being inducted into the Birmingham Hall of Fame.

U.S. MARINE CORPS

Leo D. Barlow
U.S. Marine Corps

Leo Barlow was born and raised in Henry County, Ohio, and went to Liberty Center High School. In March, 1950, at the age of twenty, he decided to go to work on the lake freighters. When he tried to sign up, however, he was turned down. He went across the street to the Marine Corps recruiting office where they were glad to have him.

Why did you pick the Marine Corps, the toughest of all the services?
Because it was across the street from the lake freighters' office.

Where did you go to boot camp?
Parris Island, South Carolina.

I've heard that Marine boot camp is really tough.
It was a rude awakening for a person from Liberty Center, with a drill instructor yelling in my face. I did well on the firing range, the best in the platoon, so the D.I. got to know me and started yelling at me by name.

Did you realize at the time that they were trying to teach you how to stay alive?
Discipline was the main thing. When you were given an order, you did it automatically, without a thought.

When you graduated, there was a great deal of pride, wasn't there?
There sure was. Being first on the firing range was a matter of pride, too.

Where did you go from Parris Island?
After three or four weeks, I finally wound up on a troop train for Camp Pendleton, California, where our division was being formed. In August, 1950, we shipped out.

Now you knew you were headed for the Far East Command*. How did you go over?
We went by ship, which took us two weeks and I got sick two days out.

The Pacific is famous for its terrible storms. Did you get into any on the way over?
No, but we did later, while we were on an LST* headed for Inchon* and that was really bad.

Where did your troopship land?
We landed in Japan, where we stayed for ten days. We didn't see anything of Japan because they were getting us ready for the landing at Inchon.

Where did you go from Japan?
We boarded LSTs and went across the Sea of Japan. As we neared the shore at Inchon, we boarded amtracks*, amphibious vehicles to go ashore.

Wasn't this a daring landing?
Yes, most were opposed to it, because of the tides, but [General] MacArthur pushed it through.

Were you being fired on as you landed and could you move off the beach?
Yes, we were in the first wave and moved in to secure a hill. We were on the offensive most of the time. The objective of the landing at Inchon was to cut off the North Korean troops, who had penetrated into South Korea, all the way down to Pusan*. The North Koreans would have to fight their way out of the trap, thus relieving the pressure on the Pusan perimeter.

You were a sergeant when you were discharged. What was your rank, when you landed at Inchon?
I was a PFC [private first class] and an assistant BAR* man, when we landed. The BAR man got wounded when we landed, so I became the BAR man. The BAR was a rapid fire 30-caliber [gun], with a 20-round magazine. We were always on the front line, supporting the machine gunners and because of our firepower, we were targeted by the enemy.

Did you have any mechanized equipment for support?
Some at times, but it was usually just infantry.

At some point, the North Koreans were supported by the Chinese. When did that happen?
That happened in October, about a month after we landed.

After landing at Inchon, did you start advancing north?
We went into Seoul and recaptured it house-by-house, and room-by-room. It

was dangerous and stressful. After Seoul, we were pulled out and made another amphibious landing. We went to the other side of the peninsula, to trap the North Koreans fleeing north from Pusan.

Were you fighting regular uniformed army units?
Some were in uniforms and some were children carrying guns or grenades. You couldn't tell who were the enemy.

Did you continue north?
We then went to hold a road junction to meet the North Koreans heading north. We became surrounded and had to call for supplies by airdrop.

Had the U.S. started using helicopters in Korea, as they did in Vietnam?
They were used to evacuate the dead and wounded, but not for troops or supplies.

When did you first encounter the Chinese?
It was in November that we first saw them at the Chosin Reservoir*. After the landings at Inchon, the UN forces drove north to Pyongyang, the capital of North Korea, and continued toward the Yalu River, which divides North Korea and China. The Chinese had threatened to enter the war and finally did as the UN forces neared their border.

It was now winter in Korea. What was the weather like and were you dressed for it?
It could get down to forty degrees below zero and we didn't have the proper gear. We lost more people to the weather than to enemy action. [Marine losses at Chosin were 836 killed and 12,000 wounded, most with frostbite injuries.] I had frostbite and lost all my toenails. I am still bothered by it today.

How were the Chinese equipped?
The Chinese just had a lot of people, some mortars, and no mechanized equipment. We had air power, which made the difference. We never would have gotten out of Chosin without the air support.

Didn't the Russians supply the Chinese with jet fighters?
That was later in the war. I left before the Russian jets came [in late December, 1950].

What was it like at Chosin?
We were at the south end of the reservoir, holding a road junction open. We were dug in and waiting. They [Chinese soldiers] would blow bugles, so we knew they were coming. They just kept coming and coming. They broke through in a few

places, but were repulsed. Some carried machine guns and some just carried a pole with a bayonet on the end and no rifle. Some carried ammunition that we found had been made in Connecticut.

Did you consider them good soldiers?
They were well disciplined, but had poor equipment and poor clothing. Their strategy was to overwhelm us with huge numbers of troops.

So now you had to withdraw?
They had roadblocks all the way back and were up on the side of the mountains, firing down on us. It was harder to withdraw than to advance. We had air power and they didn't. If they had air power, we never would have made it out.

Did most Americans get out?
Most got out but many suffered from frostbite. We came across a group of Chinese on the side of a hill and when we approached they didn't move. They were all frozen to death.

How did you do in the withdrawal?
I walked out on frozen feet. Part of the time on the road and part of the time in the mountains.

Did you get any help for your frozen feet?
No, I went to the medical station, but I saw so many bloody clothes and boots that I decide that I could still get around. When I tried to get compensation after the war, the only proof I had was a letter home to my mother, where I told her about my frozen feet.

How long did it take you to walk out of Chosin?
Probably a week to go sixty miles, because we had to fight through their roadblocks. I remember one instance, when the Chinese attacked an artillery unit. Thousands of them charged and the big guns were turned on them loaded with grapeshot. They were killed by the thousands and bodies were everywhere.

Did the Chinese make any effort to recover their dead and wounded?
They did take out their wounded, but they didn't have field hospitals like we had.

What did you do when you finally got out?
We went back to Pusan to get replacements and then we started north again.

What about your frozen feet?
I went anyway. I did get a shower in April, my first since last September.

Were you ever wounded?
No, but the man in the foxhole with me was shot in the face.

Were you there when General MacArthur was fired by President Truman?
Yes, he wanted to go into China and he made the mistake of publicly disagreeing with Truman.

How long did you stay in Korea?
I landed in September, 1951, and left in June, 1952. When I learned that my father had died, I went to Japan. After a long wait, I finally got on a ship for home and arrived in September.

Did you ever think about re-enlisting?
I spent a year in Washington and finally decided to get out. I came back to Liberty Center, where I have lived ever since.

Did you take advantage of the GI Bill of Rights*?
No, but I use the VA [Veterans Administration] health care system, which is wonderful. I couldn't afford the care they have given me.

William D. Cartledge
U.S. Marine Corps

At age seventeen, William Cartledge decided that he wanted to be a Marine, so he joined a reserve unit. After summer camp, he decided that being a Marine wasn't too bad, so in 1948, he enlisted at age eighteen. At boot camp, he soon learned that there was a world of difference between the reserves and the regular Marines.

Let's talk about boot camp.
It's a whole different world down there at Parris Island [South Carolina]. It began with the haircuts. They cut off all your hair and I mean all your hair. I had a head start, having been in the reserves, but that was both good and bad. We were taking our M1 rifles apart and instead of waiting for [the drill instructor's] instructions, I had mine all disassembled, which caused the D.I. to call me all kinds of unrepeatable names. So I learned quickly to wait for orders before doing anything.

Some men just couldn't handle the pressure and got out on what was called a section 8 discharge [a former category of discharge, for being mentally unfit]. The D.I.s enjoyed getting us in line at the mess hall, a half hour early and making us stand at parade rest, which meant that we couldn't move or talk. If we slapped a bug, the D.I. would go crazy because we had killed one of his sand fleas.

When you left Parris Island, where did you go?
I went to Camp Le Jeune in North Carolina, where I stayed for two years. When the Korean War started, I was working at the PX [retail store on a military base], because I had been a clerk at Krogers before I enlisted. I was then sent to Camp Pendleton, California. I was there about eight weeks and then went by troop ship to Japan. Being on the ocean for two weeks was not bad, but the crowded conditions were a problem. We stopped in Kobe, Japan, to drop off our unnecessary gear.

You said that you landed at Inchon*. Did you land under fire?
Yes, this was my first taste of combat. I was scared to death, as most were, but when you hear the first machine gun fire and see the first person killed in action, you realize what's happening. We got into some heavy fighting once we left the beach area and moved inland near Seoul. We fought around Seoul until we moved up to the 38th parallel*. The reason for the landing at Inchon was to cut off the North Koreans who had advanced all the way down the peninsula to Pusan*. It was a

brilliant move by [General] MacArthur. It was then decided that we would make another amphibious landing, at Wonsan* [on the North Korean east coast]. We were on LSTs* for about a week before we landed.

Let's go back to Seoul. Was that house-to-house, street-to-street fighting?
We weren't involved in that city fighting. It was the 1st and 5th Marines who did that. That kind of fighting causes serious casualties. They went door to door there and that was very tough.

So now you are in North Korea.
Yes, we went to a place called Hamhung [North Korea's second-largest city], stayed a few days, and then moved on to the Chosin Reservoir*. We went by truck for a bit and then walked the rest of the way. We met resistance along the way. Sometimes it took us an hour and sometimes days to overcome the enemy.

Did they have tanks and artillery?
They had some tanks, lots of mortars, and the troops carried automatic weapons. But we had airpower and dominated the sky at all times.

What was the Chosin Reservoir and was it important to either side?
It's just a large body of water that we passed as we chased the North Koreans north. It didn't have any strategic importance, but it gained fame as the place where we met the Chinese for the first time. We set up a command post and sent patrols out from there, but every time they went out, they met heavy enemy resistance. We finally realized that we were surrounded by Chinese. The Koreans kept telling us that there were Chinese troops, but we didn't believe them at first. One night our position was overrun by the Chinese and we lost a lot of men that night.

What did you think of the Chinese soldiers?
I'd say they were good. They could survive on far less than we could. If they had a bag of rice, a cook pot, and some ammunition, they could survive for a long time. They had inadequate clothing for the cold weather and we found many of them frozen to death. Their communications were terrible. If an order changed, they had no way to tell their forward elements. But, they had big numbers. We had about 15,000 and they had about 150,000 [estimates put the number of Chinese at Chosin closer to 250,000].

What kind of winter weather did you have?
The most severe weather was thirty-five degrees below [zero]. I don't think there is any clothing that would keep you warm in those temperatures. We wore shoepacs*. They had rubber up to the ankle and leather from there. Inside we had felt liners.

The problem was that if you moved around much, your feet would start to sweat. If you couldn't take off your boots and change your socks, you could get frostbite. People lost fingers and toes and sometimes their whole foot.

What did you do for food?
Mostly we had G.I. rations and in the villages, we could barter for root crops like potatoes and carrots.

How did you survive the cold?
I had frostbite and was wounded. We were more afraid of being wounded or killed, so we couldn't worry about frostbite. It had been decided that it was time to fight our way out of the trap. We were headed south when I was wounded and evacuated.

Were the Chinese and North Koreans right on your backs as you were headed south?
Yes, the story was that the Chinese were ordered to annihilate our division. We left California with 215 men, had seventy replacements and when we finally got out, we had twenty-one men still on their feet. Of course, they were not all killed. There were wounded and frostbite losses in that number. The cold weather may have created as many casualties as the Chinese. We were fighting our way through a roadblock, when I got hit in the leg. There was a jeep right there and our corpsman gave me some morphine and after that, I didn't feel anything. I got back to an aid station, where I got into a Navy plane, which took us to a field hospital in Hamhung.

Where did you go from there?
I was evacuated to an Army hospital in Japan and stayed about a week until I was put on a train to a Naval hospital. I stayed there for two or three weeks and was sent to a rest camp. When they decided that I was combat ready, they sent me back to Korea. It was tougher going there the second time, because I knew what to expect the second time.

Did you have a predetermined tour of duty in Korea?
They had a point system where they counted your time in combat, your time in the Corps, and if you had a Purple Heart. I was the second man rotated home from my company. I was sent to San Diego, had a leave, and was sent to Portsmouth, Virginia. I thought that I had just a few months to go, but they added on what was called the Truman Year*, which extended our enlistments by another year, because it was wartime.

Did you consider making the Marines a career?
I thought about it for a while, but I wasn't cut out for military life. Some stayed in and had successful careers, but I knew it wasn't for me.

Let's talk about the "Chosin Few," an elite organization of the survivors of the battle at Chosin Reservoir.
We are a group that formed years ago. We found each other and began to meet for reunions. We could get together and talk about our experiences and things we wouldn't normally talk about. We first met thirty years after Chosin. There were about 1,000 of us there. Although everyone has changed over the years, I could still recognize my old buddies; sometimes I could recognize them by their voices. It is a national organization, with chapters in every state.

Why is it said that there are no ex-Marines?
Once a Marine, always a Marine. It starts in boot camp with the discipline and the training. When you are given an order, you don't question it, you obey it. I think with that kind of training, there are far fewer casualties.

I've heard that you never leave your dead or wounded behind.
Yes, that is an old Marine tradition. But at Chosin, we had to leave some of our dead behind; we just couldn't get them all out, but we got all the wounded out.

Do you think that the Korean War was the "Forgotten War"?
It may have been at one time, but organizations like the Korean War Veterans Association have worked hard to ensure that it is not forgotten.

Walter A. Churchill
U.S. Marine Corps

*Walter A. Churchill Jr. joined the
Marine Corps Reserve while still a
senior in high school. He wanted to
assure his father, a Marine Corps
general, that he would not disappoint
him by joining the Navy. His father
was the commanding officer who swore
him in. Walter is pictured standing
next to a portrait of his father.*

Where did you go to boot camp?
We didn't have boot camp when I went in. I was still in school. We had our own training
and they equated two years and one hundred percent attendance with boot camp.

Did you go immediately to Korea?
No, we were at summer camp in Little Creek, Virginia [now the Naval Amphibious
Base Little Creek], when we were activated. Things were heating up, so we thought
we would be going over. When we got home in September, we were activated and
sent to Camp Pendleton, California. Forty of us boarded a ship, without rifles or
dog tags, but we got plenty of shots. We went to Kobe, Japan, and from there to
Inchon*, South Korea. When I first got there, I was a buck sergeant and my job was
to help unload ships. The invasion had already taken place. When we got ashore, I
saw my first wounded, one with a leg off and one with an arm off.

**Do you remember your commanding officers? And have you stayed in contact
with your former buddies?**
One of the first I met was Lieutenant Colonel Buzz Sawyer, a well-known Toledoan.
When I returned, I was a member of the Frozen Chosin and I am a member of the
Blue Diamond Post and a life member of the 1st Marine Division.

How did you stay in contact with your family?
They would send me cookies and one of my teachers sent me a pair of soft woolen
knit socks, which I credit for keeping me from getting frostbite. We had shoepacs*,
but we didn't get to change our socks very often. This was in 1950, so we didn't have
equipment as advanced as today. We had to sleep every night with our boots in our
sleeping bags. We didn't change clothes for a month at a time, so our socks got wet
and had to be dried out every night. We had plenty of food, although we didn't have
much variety. When they couldn't get food to us, they dropped it by air and we had
to eat it before it froze.

How close were you to the fighting?

When we were coming out of Chosin*, we combined A and C companies and had forty men, where ordinarily, we would have 500. Our commanding officer said that we were completely surrounded by 600,000 Chinese [Western estimates put 250,000+ Chinese forces at or around Chosin] so, no matter which way they go, they can't get away. We had lots of wounded, but surrendering was never an option with the Corps. Most of our officers were World War II vets who had been engaged in island fighting.

Did you ever have any R&R*?

We had our own R&R. Most of the guys got out by being shot or frozen. The gunnery sergeant and I were the only ones [who] didn't get shot or frozen. One time, I had a high fever and thought I might get to go to sick bay, but by morning, my fever was gone, so I didn't go.

Was your father worried?

Yes, he was worried, but proud and concerned about what we were doing. All of a sudden, it was his kid who was in the war, so that changed things.

When you came back, were you discharged?

I was released from active duty, but stayed in the infantry reserves. I heard about the Naval Reserve unit, in Grosse Ile, Michigan [largest island in the Detroit River, near Detroit]. They would fly to Alaska or the islands on weekends. I had a pilot's license, so I went up there to join the Marine Corps Aviation. But because I had an infantry background, I was still called on to teach classes of recruits.

Any other memories of Korea?

When we got back to Seoul from Chosin, I joined Charlie Company. The CO asked who the buck sergeant was and when I told him, he asked me if I knew anything about 3.5 rockets [M20, shoulder-fired rockets, called "Super Bazookas"]. I said that I didn't, but he said you better learn, because you're in charge of that section. That was good training, because every day when you got up, you found out who was still alive, what the mission was for that day, and you built a team from the people you had. We had a good team, so we often found ourselves assigned to an outpost in the middle of a trail or a gully, where we could meet the enemy and confront them first.

Did you get any time to relax?

If we had any time, our Chinese interpreter, who was a good cook, would make us something made of rations, sugar, salt, and spices. One time, when we were on an LST* for fifteen days from Inchon to Wonsan*, we had time to ferment fruit from the mess hall and turn it into alcohol and sometimes, our corpsmen would have medical alcohol in their canteens.

When you came home, did you go back to school?
Yes, I went to school on the GI Bill* and used it for my pilot's training.

What has been your career since the service?
I stayed in the reserves until 1977 in Marine Aviation at Grosse Ile and went to summer camp one time. I started college in 1947 and didn't graduate until 1966. The rest of my career has been in the family food business. After thirty years, I and twenty-four others retired in a very impressive ceremony at the Marine Corps barracks. I enjoyed my time in the Corps and I don't regret a moment of it.

What impact did being in the Marine Corps have on your life?
It taught me that each day, I would have to handle the problems that faced me, with the people I had.

Walter Churchill continued the Marine tradition begun by his father. His decorations include three Presidential unit citations and the Korean Service Medal with seven combat stars.

Kenneth E. Cox
U.S. Marine Corps

Kenneth Cox knew he wanted to be a Marine even before he graduated from high school. When he was still a junior at Whitmer High School in Toledo, he joined the Marine Reserves, an organization that prepares one for the regular Marine Corps.

When you went into the regular Marine Corps, you were already well trained?
Yes, some of our men went directly overseas, when they were called to active duty on August 25, 1950. The Korean War had started June 25, 1950. I got my notice on my wedding day, August 5, 1951, and I had twenty days to report for duty.

Did you go to boot camp?
Yes, I went to San Diego, where I attended communications school. After I finished school, I had to go to boot camp, because they decided that we reservists didn't get enough basic training in the reserves. After boot camp, I went to Camp Pendleton, California.

What was boot camp like?
It's something you never forget. It was twelve weeks of hell. The D.I.s were tough, but their job was to turn out Marines and teach them how to stay alive. After ten days leave and ten weeks of advanced training, I shipped out on June 19 for thirteen days on the ocean, headed for Japan. We stopped there to leave our sea bags and then continued on to Korea, which took a day. I shipped over as a rifleman, but when we landed, they asked if anyone knew anything about communications. I got placed with a mortar company as a telephone lineman to keep lines open between our company and regimental headquarters, back about five miles. I had the only bunker in Korea with two private phones. We would lose lines to storms and enemy infiltrators, which we would have to repair. We lived in bunkers in the ground, but with only ponchos for a roof. In the winter, we made fuel oil stoves to keep our bunker warm.

Why did you not have to move as our forces advanced?
We were about a quarter-mile behind the lines. Our mortars had enough range that they didn't have to move up with each advance, just change the range. At that time, peace negotiations had started, so there was not a lot of advancing, just holding of positions.

You said that you were at the foot of the Punchbowl*. What is that?
It's a valley at the base of a row of mountains, fifty miles north of the 38th parallel*
on the east coast of North Korea. We had to fire our mortars over the mountains.

**If you were sending mortar shells to the enemy positions, they must certainly
have been zeroing in on you, so you must have come under fire often. Didn't the
Koreans have lots of mortars and very few artillery pieces?**
Yes, they and the Chinese had lots of mortars and we were often under fire.

Were the U.S. Army or the South Koreans in your area?
No, it was all Marines.

Let's talk about the Korean winter.
It got to be forty degrees below, but we had the new-issue boots [see "snowpacs"]
and if we wore a couple of pairs of pants, it wasn't too bad in our bunker. We had
a mess tent, where we would get our food to take back to our bunker. So we had it
better than most.

What did you think about [General Douglas] MacArthur?
It would have been a mistake to go into China, as MacArthur wanted. So maybe
being fired by Truman was a good thing.

Did you have many infiltrators?
No, but one time we had one who threw a grenade in a jeep, causing terrible damage to
three Marines. We couldn't tell a North Korean from a South Korean, so we had to be alert.

Did you have artillery support?
They were miles back from us, but we could hear their rounds as they went overhead. We
were close enough to the coast that the ships would fire their big guns over us.

How long were you in Korea?
I stayed nine months. My enlistment was actually up while I was aboard a ship going
to Korea, but we all got the extra Truman Year* tacked on because it was wartime. I
came back to San Francisco and from there back home to Toledo. I had planned to
stay in, but I changed my mind, even though I had put in seven years.

Let's talk about the KWVA, to which you belong.
The KWVA is the Korean War Veterans Association. Our main job is to let people
know that it was a war and not a "conflict." In cooperation with the city and the
county, and a fundraiser by the KWVA, we had a Korean War memorial erected in
Toledo. It was dedicated on June 25, 2006. We also talk to children in the schools
about the war, march in parades, and meet every month.

Robert S. Darr
U.S. Marine Corps

Robert Darr was born in 1932, during the Great Depression. His family struggled, like so many others at the time. His father worked for the Works Projects Administration [WPA], a government agency formed during the [Franklin D.] Roosevelt administration to provide work for the unemployed, while his mother worked for a local department store. When Mr. Darr graduated from high school, he didn't want to follow the same career path as so many of his friends who found work in the local manufacturing plants.

Why the Marines?

I wanted to join the Air Force because I belonged to the Civil Air Patrol, but they weren't taking anyone, so I thought the Coast Guard, but they were filled up, too. When the Navy said no, I went to the Marine Corps and off I went to Parris Island [South Carolina] for ten weeks of boot camp.

Was leaving home difficult? Did you miss mom's home cooking?

Remember that I was the oldest of six kids, so leaving home wasn't that tough.

Was boot camp tough?

It was too tough for some. We had three dropouts who couldn't handle it. It was ten weeks, with tests that followed. I was sent to mechanics school at Camp LeJeune [North Carolina], for ten weeks, because I had prior knowledge of automobiles. Then I was assigned to Cherry Point [North Carolina] to the Marine Corps air station for eleven months. I had an opportunity while there, to join the MPs [military police]. I was a prisoner chaser, going to various cities to get people who went AWOL. We would take them to the prison in Norfolk, Virginia.

Where did you go from there?

After that duty, I was assigned to Korea. I went to Camp Pendleton, California, and got on a troop ship. The first few days weren't bad, but then most everybody was sick. We landed in Japan, where I spent one day. I had a chance to go into town to get a haircut and see a riot. From there, we went by boat to Inchon*, South

Korea. The invasion had taken place a year before and the North Koreans had been chased out of the South. I was assigned to a motor pool of the 5th Regiment, 1st Marine Division. I was sent to Non-Commissioned Officers School, because I was a sergeant. A Marine is first of all a rifleman, no matter what his assignment was. Then I operated a fuel dump for the regiment with two Koreans working for me.

Were there other UN troops in your vicinity?
Yes, like the Black Watch from Scotland, who we visited. They had the best Scotch whisky, which we enjoyed after duty hours. We had the opportunity to have such pleasant interludes.

Where was the main line of resistance*?
It was about 1,000 yards away. They tried shelling our fuel dump, but never hit it. One time, they hit a mess tent about 300 yards away. The guys had just come off the line and were having Sunday dinner. It was the last one they ever had.

Did the enemy have aircraft and did you come under air attack?
No, but I have seen napalm dropped on them. There were Russian MiGs* in Korea, but nowhere near us.

Did you have tanks?
We were able to use tanks but I never saw North Korean tanks.

Was it a sort of hit and run thing with the North Koreans?
While I was there, our forces would go out during the day and the Koreans would leave their observation posts and when the Americans returned at night, the Koreans would come back to the same observation post. They would trade places every few days. It was crazy. I remember that on Christmas Day, the North Koreans played Christmas songs for us.

Were you fighting an organized army?
It was not like Vietnam. The North Koreans were in a regular army, with uniforms.

What is Operation Little Switch*?
I participated in Operation Little Switch as a driver. When they released the American prisoners who were sick. . . . This happened only at Panmunjom [city along the 38th parallel* where the 1953 Korean Armistice was signed].

How long were you in Korea?
Eleven months.

Have we said all that we want to say about your time in Korea?
One of the nice things that happened in Korea was when I ran into a boyhood
friend in the chow line. I didn't know he was in Korea. He was with another
company. We spent a weekend together and then he went back on line.

Did you get any leave time while you were there?
I had two days back in Seoul on a beer run. It was a leave town for the Marines. I
stopped at a MASH* unit to visit a couple of corpsmen I knew from Cherry Point
and I went to Kimpo Air Base for supplies.

How was the food?
We had hot meals and the guys on the line would have hot food when they came
back off the line.

How were the winters?
We had a bad winter and a rainy season before that, but we had the clothes and
boots that the first guys in Korea didn't have.

Where did you go when you left Korea?
We landed at Treasure Island [former Naval Air facility in San Francisco, California]
and I was sent to Camp LeJeune [North Carolina]. I was sent to mechanics school,
but I didn't spend much time as a mechanic. I worked in an office, I was an
instructor, I was in the Military Police, and a bus driver.

**At the end of your four years, with good rank and wartime experience, were you
thinking of staying in?**
I was newly married and didn't think it was a good idea, so I opted out.

**Let's talk about the veterans organizations you are in today. After serving your
country, you continue to serve in these veterans organizations.**
I am a life member of the American Legion, Amvets, VFW [Veterans of Foreign
Wars], and served on the Veterans Service Organization for Lucas County. I am
the vice-commander of the Korean War Veterans Association, Commander of the
Joseph Diehn American Legion Post, and past county commander of the No. 45
American Legion Post.

**So you have continued to serve your country long after you left active duty.
Before we close, say a few words about serving your country for so long.**
I am chairman of the American Legion Gifts for the Yanks Program, which raises
money for veterans who are in veterans hospitals. I do this kind of work because
I like to do it, and I like the people I work with.

Did you use the GI Bill of Rights*?

I bought my first house with the GI Bill. I used the GI Bill for on-the-job training with the telephone company.

Your closing thoughts.

I am proud to have served my country and my community. One more thing—a member of our American Legion post, Roger Durbin, from Berkey, Ohio, was behind the World War II memorial along with [U.S.] Representative Marcy Kaptur. We went to Washington with some of our post members for the groundbreaking and the dedication.

Donald M. Griffith
U.S. Marine Corps

Donald Griffith was a survivor of the battle of Chosin Reservoir*. He also had the misfortune of being captured by the North Koreans and spending thirty-three months in a POW* camp. He is a member of that exclusive club, the Chosin Few.

I was born in East Toledo in 1927 and at the age of seventeen, I enlisted in the Marine Corps for four years. I was at boot camp at Parris Island [South Carolina] when the atom bombs were dropped and World War II ended. I was sent to Norfolk, Virginia, where I worked as a sentry at the main gate. From there, I was sent to California, where I boarded a ship for China. When we stopped in Hawaii, I was taken off the ship to be a guard for the Japanese prisoners still held there. From there, I went back to Norfolk to be a sergeant of the guard in 1948-49. My enlistment ran out, so I re-enlisted for two more years.

While I was in Norfolk, in the hospital, war broke out in Korea. When I was released, I went to California and boarded a ship going to Korea. After a stop in Kobe, Japan, we landed in Pusan*, South Korea. On May 29, we arrived in Seoul, where we loaded on LSTs* for Inchon*. We went in on the second wave of the invasion, while they were still shooting at us.

I should tell you that our platoon was the first integrated unit in the Marines. We had blacks, whites, Hispanics and one Polish displaced person.

After we took Inchon, we went to take Kimpo air base. Near Kimpo, we ran into enemy fire and when we went into the attack, we found that the North Koreans were hiding behind civilians. We had no choice, either kill them and the civilians, or be killed. When we were at the main line of resistance*, at the Han River, we had thirty-nine men in our platoon. We got out with ten survivors.

We secured Seoul and went back to Inchon, where I ran into a Marine I knew from Norfolk. He had a slight wound and he said I'm going to eat pork and beans for the rest of my life. He showed me the can of pork and beans that he had in his jacket that stopped a bullet from killing him. We then went to Wonsan*, where I was sent into town to find billets for our unit. As I was walking up the road, a guy yelled at me, "Where's the rest of the Marines?" I said, "Are you Bob Hope?" He said, "Yeah." He had come to put on a show and got there by helicopter before we did.

At the Chosin Reservoir*, we were surrounded by the Chinese. That's when we learned that we were going to fight in a different direction. There were 11,000 of us and 110,000 [closer to 250,000] Chinese. Ours was the last platoon to move out. When we were supposed to finally move out, the order never came. The bugles started blowing and they attacked us. A Chinese soldier hit me with his rifle butt and I was bleeding, but it was so cold that the blood was freezing.

When they captured me, they took my fur-lined parka away. They marched us for three days and three nights without any food or water, nothing. We'd fall down and grab a handful of snow to put that in our mouths for moisture, but if they caught us, they'd knock it out of our hands. They never did a thing for my wounds in all the time I was a POW. If you had to go to the bathroom, you yelled at the guard. This one night, I was really fed up and I thought, hell if they shoot me, I might be better off. I yelled to go to the bathroom and nobody answered, so I went out the door and started walking.

The next morning, I was hungrier than hell, so I went up to a house and knocked on the door and this Korean, oh, how he welcomed me. He started feeding me and he rolled me a Korean cigarette. What I didn't realize was that he had sent his son back for the soldiers, because they got paid for turning us in. About an hour later, I heard a grenade outside the house; they wanted me to know they were out there. I went out and they took my boots away from me, my snowpacs*. All I had was my ski socks on. They marched me back and made sure that the other POWs . . . saw that they had recaptured me.

I was thrown in a pigpen, where there was an American soldier, who went insane, living in the pig manure. One morning he was dead and they took him away. There was a mass grave that we put our dead in. It was never covered up, because people were dying every night. I was in the pigpen for thirty-three days before they let me out. We got transferred to Changsong, an organized POW camp, where they separated the sergeants and the officers from the other men.

Did you ever get any Red Cross packages?
Not until the prisoner exchange in Panmunjom did we get some cigarettes and razors.

Were you tortured?
Only when they recaptured me. I got down to eighty-five pounds and could hardly sit down because I had nothing but my bones to sit on.

You said that they did some medical experiments on the POWs.
Yes, I was totally blind for six weeks from an experiment. They gave us shots for bubonic plague, which made us sick for three days. They implanted a monkey gland in one guy, which eventually rotted away.

How did you make it through the winter?

We had padded uniforms and one wool blanket, and a comforter, which we sewed together to make a sleeping bag.

How did you get released?

On July 27, [1953, when the cease fire was signed], there was an exchange of POWs. We walked twenty miles to a railhead, where we got in boxcars and went to Panmunjom. We got our first Red Cross packages and were deloused, took showers, got haircuts, shaved, and sent telegrams home. The Chinese and North Korean POWs were taking off their American clothes and throwing them away. I wish that I'd had clothes like that when I was in prison. I got a letter from my younger brother, who told me he was in the Marine Corps at Cherry Point [North Carolina]. We got to spend thirty days together, before he went to Korea. After I was home, I got a letter from President [Ronald] Reagan, inviting me and the other POWs to a ceremony at the White House to receive a POW medal. It was the greatest honor I ever received.

I have a card that says that you are a member of the Chosin Few.

That is an exclusive fraternity of honor. You had to be at Chosin Reservoir to be a member.

This has been a remarkable interview and it is remarkable that you are here to tell it.

I told my wife that every day we have together is a bonus day and I've had many bonus days, because, I should have died ten times in Korea.

Robert F. Hassen
U.S. Marine Corps

Robert Hassen went with his friend, Maynard, into the Marine recruiting station, after they were turned down by the Navy. He didn't want to join the Marines but Maynard did. Somehow he got talked into signing up with Maynard. But then Maynard failed the physical and went home and Robert found himself in the Marine Corps.

I went to Parris Island [South Carolina] for boot camp, with all its bugs, and then to Camp Lejeune [North Carolina], where as part of the 2nd Division, we took a six-month cruise to the Mediterranean Sea, where we saw Greece, Italy, France, and French Morocco. I then joined an [ammunition] company. We loaded 155mm ammo on a train. It was hard work, but not for a farm boy. We shipped out with the ammo to Kobe, Japan. We supervised the unloading of the ammo by Japanese workers. After two weeks, we boarded LSTs* for Inchon*, South Korea. We went to Seoul and then headed north, above the 38th parallel*, on a narrow gauge railroad train, to North Korea. We lived in tents, although it was twenty below zero. We didn't have all the winter gear, but the Army did, so we stole it from them.

In November [1950], the Chinese crossed the Yalu River* into Korea and attacked. There was said to be one and a half million Chinese [Western estimates put the initial Chinese forces entering Korea at 270,000] and all we had was one division of Marines and half of an Army division. Many of [the Chinese] had no weapons, but if the man in front of them was hit, they would pick up his weapon. One night as we were guarding our ammo dump, they attacked. They were all dressed in white clothes, so we couldn't see them against the snow. One of them threw a hand grenade that hit my back. I rolled over to throw it back, but couldn't find it, so I rolled away, but got hit with shrapnel in my arm. I went to a field hospital and then was flown to a Naval hospital in Japan. I was in a rehab facility near Kyoto for five months. They couldn't remove the shrapnel, so it's still there. They sent me back to the States in May, 1951. I went to Norfolk, Virginia, where I pulled guard duty, until I was discharged in August, 1952. So I came home with frostbite and an injured elbow. I get disability for both.

Back to Korea. Did your unit ever cross over into China?
No, we never crossed over into China. [General] MacArthur wanted to keep going into China, so President Truman fired him, so he wouldn't start World War III.

Were you there when the Americans were surrounded at Chosin*?
No, I was already wounded and in Japan.

While you were in Japan, did you ever get into the city?
My arm was in a sling, but other than that, I was OK, so I had a chance to travel on the Japanese trains. The signs were all in Japanese, so we never knew where we were going. I went to Tokyo and got to see the Emperor's palace. While I was there, I ran into an old school friend who was in the Navy. What are the chances of seeing an old friend thousands and thousands of miles from home?

How were you treated by the Japanese?
They respected the Marines because they knew that they had beaten the Japanese in the island fighting.

How long were you in Japan?
From December 1 to May 25 [1951].

Do you feel that the U.S. government was grateful for your service?
Yes, other than fighting the VA [Veterans Administration] doctors to get disability for my frostbite.

Did you use the GI Bill*?
No, but when I came back, my old employer didn't want to give me my job back. That was illegal, so they soon changed their mind and took me back.

Do you keep in touch with old buddies?
Yes, two of them have stopped at the house and spent a few days with us. I belong to the DAV [Disabled American Veterans], the VFW [Veterans of Foreign Wars], and the Catholic War Veterans.

James L. Hays
U.S. Marine Corps

James Hays was a child of the Great Depression. His father worked for the Works Progress Administration and had a couple of factory jobs to make ends meet. Mr. Hays and some of his friends from Toledo's North End who went to Woodward High School joined the Marine Corps Reserve. They not only got to wear uniforms, but also got $3.00 per month.

I was at the Marine Reserve summer camp in Little Creek, Virginia, when the Korean War broke out on June 25, 1950. I was only seventeen at the time. When I was eighteen, I was activated and sent to Parris Island [South Carolina] and then to Camp Pendleton, California. In June, 1951, I went to Korea on a troop ship. It took us eighteen days. We landed at Pusan* and flew up towards the line; we rode on trucks part of the way and then walked in to the line. I spent one year in Korea; most of that time was in combat.

What major battles were you in?
We were in the Punchbowl* area. While the Army was at Heartbreak Ridge* and Porkchop Hill*, we were on Hill 749. They named their hills, but the Marines numbered theirs. I was at Hill 749, Hill 812, and Hill 884. We were supposed to go to Hill 1052, but we took so many casualties, we couldn't go on. We went forty-eight straight days in combat and then got two weeks rest, showers, and clean clothes. On the day before Christmas [1951], we had a hot meal and then relieved the guys on the hill, so that they could have a hot meal on Christmas Day. We were back on the hill for thirty more days. It was thirty below zero on Christmas Eve.

How did you get food on the line?
I had three hot meals in a year. The rest of the time it was C rations.

Did you make any good friends in Korea?
Kenneth Tolle was a friend from Woodward High School. He was wounded in action and lost both legs. He recently died.

Were you ever wounded in combat?
After all those months in combat, it's a miracle that I was never wounded.

Were you able to communicate with your family?
We could write and receive letters. One time, my mother was in the hospital and she complained that I didn't write to her often enough. The Red Cross heard about it and their representative in Korea came to division headquarters to look into it. They sent a jeep to take me to HQ to interview me. They said my mother was sick and wanted me home. They asked if I wanted to go home, but I said no, that I would wait until my time was up in Korea. So I stayed until my year was up.

Did you have any leave time while you were in Korea?
We had no leave and no days off in a year.

Where did you live?
We lived in bunkers all the time. We didn't have sand bags, so we used logs to reinforce our bunkers and for the roof.

Did any entertainers come to Korea?
Betty Hutton [an American actress, singer, and comedienne] came once. We had to walk five miles back to see her. She's the only one I ever saw.

When did you come home?
I came home in June, 1952, when my year was up. I wasn't discharged until March, 1953, when my four years were up.

What medals did you earn in Korea?
The United Nations Medal, Korean Service Medal with one battle star, and a Korean government medal. The Marine Corps awards very few medals. I also have all the letters that I received from my mother and my girlfriend and all the letters that I sent to them.

What did you do when you came home?
I worked for Libbey Glass for six months, until I got a chance to get into the electrical trade. I spent the next forty-two years as an electrician. We got married in 1954, had four children, and five grandchildren. I got a GI loan* to buy the house we lived in for forty-two years, until we moved to Curtice.

I stayed in contact with my friend, Kenneth Tolle. He was in the hospital in Chicago, when I took him a hand-operated car from Roth Pontiac. I took the train back to Toledo. I went to Philadelphia to see him again. One issue of Leatherneck magazine, the Marine Corps magazine, had a picture of his helmet sitting on his upturned rifle. He died a few years ago.

Edward J. Kusina
U.S. Marine Corps

Edward Kusina didn't want to wait to be drafted in the Army, so he and a friend joined the Marine Corps. They were sent to Parris Island [South Carolina] for boot camp and after they had their heads shaved, they said, "What in the hell did we do this for?"

After boot camp, I was sent to Camp Lejeune [North Carolina] to driving school. After I washed out of there, I was a prison chaser and then an Admiral's orderly. In June, 1950, the Korean War began and I was sent to Camp Pendleton, California, where I became part of a machine gun squad. I was in the invasion at Inchon*, to cut off the North Koreans, who had penetrated all the way down to Pusan*. Our landing craft got stuck in the mud, so we had to get out and wade ashore. We got ashore under fire and wandered around trying to form up our unit in the dark. We got into a bitter fight on the way to Seoul. In Seoul, it was house-to-house fighting.

How long did it take to get through Seoul?
About seven to ten days. We went down the main street following the tanks.

Were the North Koreans like a regular army, dressed in uniforms?
No, the only way we could tell them from the civilians was that they carried guns. It was like a guerilla war. They would hide among the civilians and we couldn't tell them apart. It was chaos; they were in the ditches, in the houses and in the mountains. They were everywhere.

Did they have any other weapons other than the rifles they carried?
They didn't have any mortars or tanks, just rifles.

What kind of a city was Seoul?
It was a big modern city, with multi-story buildings, but the North Koreans didn't occupy the upper floors. They fought from the main floors. They just wanted to get out of there; they didn't want to hold the city. After Seoul, we were pulled back to Inchon and then by boat to Pusan to reorganize. I ran into eight guys from Toledo I went to school with. They were just coming in. Their reserve unit in Toledo had been activated.

Where did you go next?
We got on boats so far and then on trucks to take us to Chosin Reservoir*, up near the Chinese border. That was in November [1950] . . . the Chinese [had] entered the war. They were dressed in quilted uniforms and those without rifles carried a

long pole with a steel rod on the end. If the man in front of him was shot, [he] would pick up his rifle. The Chinese didn't have much regard for their people. They charged and just kept coming into our machine-gun fire. They would pile up in front of our gun. Life was cheap there.

How did you get out of Chosin?

We walked out for thirty-six hours. We wouldn't get on the trucks, because if they stalled, they were shoved off the mountain, maybe with us in it sleeping. We blew up all of our equipment at Chosin, trucks, bulldozers, and tanks. We held on to the chain hanging from the back of the truck to help us walk. Some guys got so tired that they just lay down on the side of the road in their sleeping bags and wouldn't move. We kicked them and told them to get up or be captured. They wouldn't or couldn't move. They had given up. I kept going, because the thing I feared most was being captured. We leap-frogged down with other units. We would pass through them and then we would wait while they passed through us. We were under fire, but we had the Marine Corsair [Douglas A-1 Skyraider] planes dropping napalm, to hold off the Chinese. We got on boats to go out to the ship that would get us out of there. Our boat started to sink, so they took [us] on board ship first and got a bunk.

Were American losses heavy at Chosin?

I know that we lost four men out of our ten-man machine gun squad.

You have two Purple Hearts.

The first one, I was hit in the stomach. They patched me up and sent me back to my unit. The second one was for the mortar shrapnel that I got at Chosin. Fortunately, it wasn't bad enough to keep me from walking out. There was an announcement, after we got back from Chosin, that if you had a Silver Star, a Bronze Star, or two Purple Hearts, you were going home. I gave away all my equipment, including my rifle, but as we were going to the ship, we got ambushed, and I didn't have a weapon. We went to Japan [in April, 1951] and sailed to San Francisco.

What do you think about serving your country?

I did what I had to do and I'm proud to be a Marine. But I can still see those guys sitting on the side of the road as we were leaving Chosin. We didn't abandon them, but we couldn't take them with us if they wouldn't walk.

What did you do when you got out?

I joined the Toledo Police Dept, where I worked for thirty-one years.

Mr. Kusina has a case, made by his daughter, with all his medals, stripes, his arm patch, his service ribbons, and a picture of her father as a young Marine.

Donald Millington
U.S. Marine Corps

*In 1950, when Donald Millington graduated from
high school, he wanted to join the Marine Corps, but
his parents made him wait until he was eighteen. In
February, 1951, he enlisted. He said he knew he wanted
to be a Marine when he saw a relative wearing his
Marine Corps dress blues.*

The first night, I was terrified. I made some mistakes
and got chewed out for it. At the end of our training,
the D.I.s even sort of liked us. I realized that they
were trying to teach us how to stay alive. Boot camp was ten weeks of Marine Corps
history, weapons, discipline, and drilling. I came out a PFC [private first class] and
went to Camp Pendleton, California, where I worked in a supply depot for a year.
I asked for a hearing to request a transfer to Korea. My request was granted and I
took six more weeks of additional infantry training.

We sailed for Yokohama, Japan on a troopship, which was tedious. When the
guys got seasick, they threw up everywhere and the odors were terrible. I enlisted
with a guy from home, but hadn't seen him in fourteen months. I was surprised to
see him on the ship and sleeping next to me. From Yokohama, we went to Inchon*
to be assigned. By this time I was a buck sergeant, but had no combat experience,
so they sent me to an eight-week infantry combat leadership school. At this stage of
the war, there was a shortage of experienced non-coms [non-commissioned officers],
so they had to train a new group. After completing that school, I was assigned to an
infantry company. I felt a lot better after the training about leading men that had
been there and had experienced combat. I was assigned to Fox Company.

How many men did you have?
Initially I had a squad of thirteen men and later, a platoon of forty men.

Where were you stationed?
We were near Panmunjom*, the truce city. Next to us were the Black Watch Scots
and later a Turkish unit.

Was there a battle front?
Yes, about 600 yards away, we faced the Chinese who were good fighters. We had
trenches and bunkers and outposts out in front of that. Behind us would be the mortars
and further back the artillery. The Chinese had about the same arrangement.

Did you have tanks?
Yes, but they were not too useful in the terrain of Korea.

Were you constantly under fire?
Some days it would be constant and some days there would be none at all. The Chinese were very good with mortars. One day, three men lay on top of their bunker to get some sun and one mortar round killed them all.

Did you dominate in the air?
The Russian MiGs* were at first the best plane in Korea, but later the Americans dominated the air with better planes and more of them.

Did the Chinese attack often?
No, it was mostly patrols in no man's land that fought. We would have patrols for different purposes. Some were ambush patrols to engage the enemy, some were to probe and attack, and some were recon patrols.

During this stalemate period*, you got wounded. How did that happen?
Another sergeant and I took a relief unit out to the outpost and while we were there, the Chinese hit us with mortar fire. We were in a bunker, but got out to check on the men in the other bunkers. A mortar round fell between us and we were both hit. Corpsmen came out with stretchers, under fire, and got us back to the battalion aid station.

I went from there to a hospital ship, to Japan, Hawaii, and California. I had lost an eye, my right hand was damaged, and I had shrapnel in my head. The other sergeant got hit in the back, but he was wearing a flak jacket, which saved his life. We both got back to Great Lakes [Naval Station] in Chicago. I was discharged and given a stipend for life from the Veterans Administration.

Do you now use the VA health system?
The people at the VA in Toledo and at the hospital in Ann Arbor are outstanding.

Did you use the GI Bill* of Rights?
I used the GI Bill of Rights to go to college and to buy a house.

Donald Millington has a chest full of medals: Purple Heart, Combat Action Ribbon for service in Korea, Navy Unit Commendation to the 1st Marine Division, Good Conduct medal, National Defense Service Medal, Korean Service medal with two Bronze Stars, Republic of Korea Presidential Unit Citation to the 1st Marine division, United Nations Service Medal, Republic of Korea War Service Medal, and the Rifle Marksman Badge.

He has been a member of the Military Order of the Purple Heart, Disabled American Veterans, Marine Corps League, Korean War Veterans Association, and the 1st Marine Division Association.

Don J. Mooney
U.S. Marine Corps

Don Mooney joined the Marine Reserves in June, 1948, so that he could choose his branch of service, instead of waiting to be drafted. When the Korean War began, his unit was immediately activated and sent to Camp Pendleton, California.

At Pendleton, we got our shots, our gear, and our unit assignments. We then got on troop ships, for a rough trip to Kobe, Japan, where we picked up our combat gear. We went in on the second wave at Inchon*. Our job was to cut off the North Koreans, who had penetrated all the way down to Pusan*. After routing the North Korean forces, we got back on boats to take us to the east side of Korea. We bobbed around for a week on LSTs* while the harbor was cleared of mines. We finally landed and started north to Chosin*. We were originally supposed to stop at the 38th parallel*, which separates North and South Korea, but General MacArthur decided to go further north to Chosin and the Yalu River*, which divides North Korea and China. It was late [October 1950] that the Chinese crossed the border and attacked us.

I was in an 81mm mortar squad, so we were about 500 yards from the main line of resistance*. Normally, we would move our position as the MLR changed, but the ground was frozen so hard and so deep that we couldn't dig in to set our mortar. It was about forty-five degrees below zero and although they didn't have wind chill calculations back then, it was later calculated that the wind chill factor was 115 degrees below zero. We had parkas and shoepacs* left over from World War II, so there were many cases of frostbite. I don't know what elevation we were at [3,400 up to 4,500 feet, depending on location], but we were above the clouds and the air was very thin, which made breathing difficult. We were easily exhausted from working in that thin air.

Our mission had been to secure North Korea, but when the Chinese entered the war, we had to reconsider, because there were eight Chinese for every American. I was wounded on the first day of our march out of Chosin. Navy corpsmen came out with a litter to take me to an aid station. They were under sniper fire all the way and when the sniper fire got too close, they would drop the litter and position themselves to protect me from the snipers. I'm not sure I could do that, but they were very special people. While I was on a stretcher at the aid station, we came under fire from snipers and I was shot again. They also killed a doctor and shot down an evacuation helicopter.

After I became one of the more seriously wounded, they took me to the airstrip we had built and I went out by plane. When they got me to a field hospital, they operated on my stomach wound and as I was being taken from that hospital, it, too, was under fire. I was taken to a hospital ship, where they operated on my head wound. Then I went to Japan and from there, I was flown home by MATS, the Military Air Transport Service, with a stop at Hickham Field, in Hawaii. I went to the Mare Island, California hospital, where we were greeted by Hollywood celebrities. We were able to call home from there, with the charges paid by the celebrities. I spent thirteen months at Bethesda Naval Hospital, Maryland, undergoing plastic surgery and then I was separated from the Marine Corps.

Let's go back to Chosin. Were you overwhelmed by the Chinese?
We were surrounded. We had air drops to bring us food and ammunition.
All of the food was frozen, as were our C rations. There were times that the Marine units would carry out missions for the sole purpose of recovering their wounded and dead.

Were these regular Chinese soldiers?
Yes, they would come at us like kamikazes and their dead would pile up in front of us.

Did you have air superiority?
The Marine Corsairs* would give us close air support and the jets would bomb targets like bridges and roads.

Were the Russian MiGs* a presence?
Our planes had annihilated them, because their pilots were so poorly trained.

What about tanks?
We had a few tanks, but the roads were too narrow for tanks. They had been built for oxcarts. As we were heading south, with the Chinese harassing us, Chinese soldiers would surrender to us, although they had superior numbers. They had very little food and poor clothing. They wore tennis shoes and they would strip dead Marines for their clothes, particularly their boots. They carried maggots with them in jars and if they were wounded, they would put the maggots on the wound to keep it clean.

When [General] MacArthur exceeded his orders and was fired, did that have an effect on the Marines strategy?
I don't think the Marines paid that much attention to MacArthur. They knew the situation and knew how to handle it.

Don Mooney's service to his country is reflected in the medals he has earned: the Purple Heart, Combat Action Ribbon, Presidential Unit Citation with Bronze Star, Good Conduct Medal, National Defense Service Medal, Korean Service Medal with two Bronze Stars, Republic of Korea Presidential Unit Citation, United Nations Service Medal, Republic of Korea War Service Medal, and Rifle Marksman Badge. He continues to serve his country as a member of the Marine Corps League, which provides burial service for deceased former Marines. He supports the Make-A-Wish Foundation, and has formed the Young Marines of Toledo organization to teach young men discipline and Marine principles.

Daniel C. Seeman
U.S. Marine Corps

To the extent that you know yourself and your enemy, you will win most battles. To the extent that you know yourself and don't know your enemy, you will win half your battles. To the extent that you don't know yourself and don't know your enemy, you will lose them all. So said Daniel Seeman, former Marine colonel and retired University of Toledo professor, during his interview for the Veterans History Project. Those words were first written by Sun Zhu, an ancient Chinese military general and strategist, and are among Mr. Seeman's words to live by.

Let's begin in the beginning, before your military service.
I was born in Walbridge, Ohio and moved to East Toledo, where I went to Waite High School. I got a scholarship to Columbia University, to play basketball and football. In July, 1952, I joined the U.S. Marine Corps. After boot camp, I went into Office Candidate School to become an officer. It was an intense ten-week course at Quantico, Virginia. I then went to Basic School in 1952. One of my classmates was Adlai Stevenson, whose father was the governor of Illinois who went on to become a senator [and the Democratic candidate for President in 1952 and 1956].

Did you go on to flight school?
I tried to get into flight school, but didn't make it because of my eyes. I wanted to stay in the air side of the Marines, so I went to Air Defense Controller School at Cherry Point [North Carolina]. After that, I went to El Toro, California, to Marine Air Controller Squadron No. 4. While I was there, I played basketball for the base. As soon as basketball season was over, in February, I was sent to Korea.

What did you do as air defense controllers in Korea?
Our task was to scramble planes to go shoot the enemy down. It was all done with radar. I remember an instance when a pilot was lost. He thought he was up north, but he was down south and running out of fuel. Luckily I was able to find him and get him back before he ran out. And I remember a bad day, when we lost two F86s* that crashed into the Yellow Sea. We were located on a mountain and supported the Air Force B25s [World War II bomber, built by North American Aviation], which would go on missions to North Korea. We would guide them in on radar and when the mission was over, we would guide them back. We covered all the aircraft for

southwest Korea. If our radar detected aircraft coming in, we would scramble our planes to see who they were. It was an important experience for me because we were responsible for the lives of all those pilots.

What was the most serious thing that happened to you?
Probably saving the lost pilot in Korea. Another was time that I pulled the young kid out of the ocean and saved his life. It was too rough that day, but he went out anyway. He was about one hundred yards away and drowning. I just barely got to him and brought him in.

What did you learn in the military that had an impact on your civilian life?
I learned about dealing with people. In our unit in Korea, I saw how morale would change for better or worse, depending on whether a new CO [commanding officer] was an autocratic or a democratic leader. My doctoral dissertation was based on that subject. And I have done a lot of leadership seminars over the years. I learned that you are only as good as the least person in your unit.

When did you get your Ph.D.?
Long after I left the service. I was in business for about ten years, when I decided to go back to school for a masters degree. I got that in psychology in 1960. I started teaching at Com-Tech [the former University of Toledo Community and Technical College] and I have been at the university [of Toledo] for forty-five years. I was an assistant dean and director of student activities, which was the most important job that I had at the university, because I began mentoring twenty to twenty-five students who were Vietnam veterans. I think that over the years, I have taught 20,000 students.

Paul D. Smith
U.S. Marine Corps

Paul Smith was born in Pemberville, Ohio, in 1931 and joined the Marine Corps in 1951. He went to boot camp at Parris Island [South Carolina]. After boot camp, he was a rifle coach for a year. He was sent to Camp Pendleton, California, for advanced training, combat training, and cold weather training and from there, he went to Korea.

Where did you go in Korea?
We went to Inchon* and from there to a location just north of the 38th parallel*. The Marines had been up north to Chosin*, but I never got that far. We were in places called Vegas, Reno, and Carson. They were all outposts and that's where I got wounded on March 28, 1953.

What was the objective of an outpost?
Advanced warning of enemy attacks. When I got wounded, the enemy had an outpost on higher ground.

Did you have air support and tanks?
I don't remember. All I know is that I was trying to save my butt. We had been in reserve, but there was a battle and it was now our turn to go to the MLR*. I got wounded in the right arm, right elbow, and right hip. If it had not been for my flak jacket, I wouldn't be here today. I was digging the hole deeper, when I was hit. If I had been in the hole, I would have been killed. There were three Paul Smiths in D Company, with one in each squad. The sergeant said that I was in the wrong squad and replaced me with another Paul Smith, who was killed on a patrol.

Tell us about the Korean winter?
When [the troops] first got there, they weren't ready for anything. We were never told that the winters would be so cold. The clothing was all right. We had what we called Mickey Mouse boots* [see "shoepacs"]. They were great, but if you moved around too much, your feet would sweat and when you stopped, your feet would freeze. If I was going on patrol, I would never wear them. If I was going to sit in an ambush, I would wear them.

Where were the outposts?

It varied. One was two hundred yards ahead of the MLR and there were two beyond that. They were listening posts; I didn't like those. Two or three men would be 200 or 300 yards out there all by [themselves]. Most everything we did was at night, like patrols and ambushes.

Explain what a Marine patrol is.

A patrol is a squad with a machine gunner, who has two or three men to carry ammo and the tripod, and a corpsman. The object of a patrol is to see what the enemy is doing. We were told to keep the man ahead of us just far enough away that we could still see him. One night, we were on patrol and I was following the man in front of me, when he stopped, so I stopped and the men behind me stopped. The man in front told me that he had lost the man in front of him. So there we were out there, not knowing where the patrol went. We continued on until we came to a fork in the trail. We took the wrong trail and got into high brush, where the Chinese could lie in wait for us. The Chinese would hide on each side of the trail and would grab a man and start running with him. We were scared of being taken prisoner by the Chinese, so we stopped. We got back to our lines, but I told the guy who had been ahead of mc that I was never going to have him in front of me again.

What is a recon?

A recon is a group of volunteers who go out in front to harass the enemy.

What was your relationship with the South Korean soldiers?

They were good soldiers. They wouldn't wear flak jackets because they said that they were too noisy. They would go out on night raids against the Chinese. They were very quiet and they always got a prisoner or two.

What about the Chinese and North Koreans?

They always attacked in a mass and tried to overrun you to get into hand-to-hand combat. Some Marines experienced that, but our outfit never did.

Tell us about getting wounded.

I was scared to death. There was a big boom and I didn't know if I was still there. If anyone tells you he wasn't scared in combat, he wasn't there or he's lying like hell. You're scared, but not paralyzed, because you know you have to defend yourself. When you're in a battle, you can't stop to think about it, you have to react, without thinking. That's what the Marines teach you.

You were wounded very severely, weren't you?

They took me back to a half track [vehicle with regular wheels in front and a caterpillar track in back] and back to an aid station. A helicopter had room for me and we went to another aid station. From there we were put on a big helicopter and taken out to a Danish hospital ship. I was transferred to an American ship in Japan. I was drugged. So I don't remember much about the trip or the surgery.

You showed me a telegram that was sent to your mother. She must have been really scared.

She got a letter from the Marine Commandant, telling her that I was wounded, but getting the best of care. It was another thirty days before she got another letter, telling her that I was all right. A doctor told me my arm was broken and when it healed, I would go back to Korea. But a corpsman said, "Don't worry, with that shattered elbow, you're not going back to Korea." I was taken before an evaluation board in Japan that thought it best to send me back to the States. I finally got put on a plane that took us to Hawaii and continued on to Mare Island [Naval Shipyard], California. I was sent to Great Lakes Naval Hospital [Chicago]. I got a weekend pass and went home to Toledo.

Did you then get discharged?

Yes, It was called temporary retirement and I was given a sixty percent disability.

Paul Smith continues to serve his country through his work with veterans organizations. He visits schools every year to tell the children about the Korean War and the people who served their country.

Robert E. Smithers
U.S. Marine Corps

*Robert Smithers was told when he first entered
the Marine Corps: Do the damn best you can.
Don't volunteer for anything, but if you're
picked, do your duty and make yourself proud.*

*Robert Smithers went on his honeymoon
and came home to a draft notice. He was sent
to Parris Island [South Carolina] to become
a Marine. After boot camp and a ten-day
leave, he went to Camp Pendleton, California,
for training as a gunner in a 3.5-inch rocket
launcher outfit.*

We left from San Diego on a troop ship, where the first three days were hell on
earth. It was worse than nine and a half weeks of boot camp. It helped that I was
told to start eating and keep my stomach full. I was down in the bottom of the
ship in a sea of humanity and the smells were horrible. We landed at Inchon* and
were moved to the front and divided up into different units. I was assigned to H
Company, 3rd Battalion, 5th Marines. We spent a few days of orientation behind
the lines, before we joined our unit on the MLR*, where we would spend three to
four weeks at a time on the line. We were issued flak jackets, which we wore every
waking moment, even when in the head [military toilet]. You quickly learn to
accept that there is no running water, no flush toilets, and no home-cooked meals.

Were you near the 38th parallel*?
Our MLR was the 38th, which varied by as much as 600 yards and in between was
no-man's-land where the patrols went out. I didn't go on patrols; that was done by
rifle companies.

What was your job on the line?
There were few times when I actually fired my rocket at a target—once at a tank
and another time at a machine-gun position. A rocket can only be fired twice
from one position, because we draw enemy fire. Nobody wanted us or tanks around
because we drew enemy fire. When a man on the line was called to go on patrol, I
would fill in for him. One time a patrol went out and was surrounded. All forty-one
were killed and the lieutenant was propped up so we could see him. We went out to
retrieve the bodies, in the Marine tradition of never leaving their dead behind.

What was the terrain like?
It was mountainous. We didn't have tanks, unless we called them up and then they would go back behind the lines, because they were too good a target. We had Marine F84s* and F86s* that would drop napalm or bombs on their lines.

What was the weather like?
Where we were, the weather was much like back home.

Did you have any leave time?
I had some time in Japan, which I spent with a friend.

How were you treated in Japan?
There was no animosity, but when we arrived, we were surrounded by a bunch of kids. Their jobs were to mark our clothes with a chalk mark. The chalk mark was to alert the merchants that they were dealing with GIs on leave, so they wouldn't have to negotiate prices with them. The GI on leave was not going to find any bargains, with a chalk mark on his clothes.

How long were you in Korea?
I was there eleven months. When I got back to the States, I spent the last five months of my time at Camp Lejeune [North Carolina], doing things to kill time. Once, although we were short timers, we were ordered to take a familiarization ride on a C118 plane, called the Flying Boxcar. It had a bad record of crashes, but we got back safely.

What did you do when you got back home?
I came home to my wife and child and went back to my employer, Libbey-Owens-Ford. They honored their commitment to give returning veterans their old jobs back.

Do you have any interest in going back to Korea and/or Japan?
I have no interest, but if someone were to plan a group trip, I might consider it.

Do you belong to any veterans groups?
The 5th Regiment has annual reunions. I attended when they were in Indianapolis. H Company, 3rd Battalion, 5th Regiment has its own newspaper. They report on the latest news, obituaries, and ask that we help them search for veterans with whom they have lost contact. It is published three or four times a year. So there is still lots of camaraderie, after all these years.

Richard S. Thompson
U.S. Marine Corps

"After we were deloused and shaved bald, we stood at attention for three hours in the nude. Then we started running and they threw clothes at us. We were now to be pissants and from then on, everything was designed to build a Marine."

Such was Richard Thompson's introduction to the Marine Corps. In spite of the months of humiliation, he came out never forgetting the experience and always proud to be a Marine.

Why did you choose to join the Marines?
My buddy and I decided that we wouldn't wait to be drafted, so we went to join the Navy. They were full, so my buddy said that we should join the Marines. I thought he was crazy, but when asked if I were man enough for the Marines, I had to accept the challenge. We had seen a John Wayne movie that showed how glamorous the Marine Corps was and were shocked when we got off the train at Parris Island [South Carolina].

How long was boot camp?
Ten weeks. It was a proud day when we finally graduated.

Did the D.I. [drill instructor] really get right up in your face?
Oh yes, but he never touched us.

Where did you go from Parris Island?
I went home for two weeks. It was at that time that General MacArthur was being relieved of his command. About May 1, [1951] I went to Camp Pendleton [California], for combat training. The training was superb. One important thing we learned was that you never leave a buddy on the battlefield. That will never, ever leave me.

Where did you go next?
We got on a terrible tub of a troopship. There were 5,000 of us with guys being sick all over the place. All we did was hurry up and wait and stand in chow lines. We ate standing up because the ship was rocking back and forth. We could wash, but we didn't take showers, because the water was so bad. We slept nose to nose in bunks three high. Thank goodness we were eighteen years old and could stand it. When we got to Pusan*, they called us into the hold of the ship and issued our ammunition. That was a scary day.

When was that?
It was August, 1951. I was sent to a POW camp in the southern part of Korea
and was assigned to the Headquarters Battalion. I ran into ten Marines, who had
been at Inchon* and up to the Chosin Reservoir*, in November, 1950. They said
that their one division was surrounded by six divisions of Chinese and they had to
walk seventy-five miles to get out. It was the first time that the Marines had ever
retreated. Then, in the second week of November, we were sent north of Seoul. I
remember that Seoul was almost completely bombed out, like Berlin in 1945.

Let's go back to when you arrived in Korea. What were your first impressions?
It was barren, as you saw on the television show, M*A*S*H. Most of what I saw was
rice paddies and the little huts that the people lived in.

And what about the people?
I remember the kids. I brought back the faces of the kids. How happy they were
when we gave them a stick of gum.

When did you go home?
I went home in May, 1952. I brought home no scars. I never killed anyone. I never
saw anyone killed. I'm not sure how I would have handled that. Going back on
the ship wasn't as bad as going over. I remember the thrill of docking at San Diego,
as the Marine Corps band played. As I walked down the gangplank, a Red Cross
person handed me a carton of milk, which I hadn't had since being in Korea. When
I touched American soil, I decided that I would never complain again.

Richard Thompson came home, used the GI Bill of Rights to go to the University of
Toledo and became a schoolteacher. It was obvious throughout our discussion that he was
honored to be a Marine and grateful for the life that was given him.*

ACTIVE RESERVES

When World War II ended in 1945, the veterans returned home, put their uniforms in mothballs, and got back to civilian life. But just five years later, America was in another war, in a far off, little-known country—Korea. Those who were in the active reserves, the National Guard, or still in one of the services, were again called to serve their country. It is important to recognize those who twice answered the call to duty.

Karl Alberti
U.S. Army Air Force

Karl Alberti and his family had suffered through the long years of the Depression and were just recovering when World War II began for America in December, 1941. Karl wanted to do his part to defend his country, so he went to join the Marines, but they wouldn't take him. So he went to the Navy—same answer. But the Army said that they would be glad to take him. After basic training, he went to school to become an ordnance expert. He learned about all the Army's weapons and went on the Queen Elizabeth [ocean liner operated by the Cunard Line from 1938-1969] to the European Theater of Operations. He spent time in Scotland and England working on aircraft armor, and at a Royal Air Force base, in Atcham, England, before he was sent to Chantilly, France, in November, 1944.

He returned to the States for a thirty-day R&R*, then was sent to Walla Walla, Washington, to be redeployed to the Pacific Theater as part of a missile squad, in anticipation of the invasion of Japan. When the atomic bombs were dropped and the war ended, he was discharged.

In 1952, there was a need for explosive ordnance disposal experts, so reservist sergeant and ordnance expert Karl Alberti was activated as a second lieutenant. He spent six months teaching ordnance disposal until the fighting ended in Korea. He was offered a post in the regular Army, but he chose to retire from the service for the second, and he hoped the last, time.

Eugene F. Eversole
U.S. Army

Eugene Eversole was a career man. He served in World War II, Korea, and Vietnam. He served from August, 1943, to January, 1971. His basic training was at Camp Claiborne, Louisiana, where he learned to be an infantryman and an engineer in a port construction group. His unit went to the European Theater of Operations on the Queen Mary [ocean liner run by the Cunard Line from 1936-67]. His first stop was Scotland and then he went to Normandy, in June, 1944 on D Day+7. The conditions were still terrible. The dead and wounded had been removed from the beaches, but there were still dead animals and wrecked vehicles to be seen. Cherbourg [France] was the only large port available to the Allies who needed to bring in shiploads of supplies daily for the war effort, but it had been destroyed by the Germans. Mr. Eversole's port construction unit put Cherbourg back in working order in a few months.

He left the engineers in March, 1945, and earned a Combat Infantry Badge* for being in action with the infantry. On VE Day [Victory in Europe], he was with the 66th Infantry Division in Germany. In January, 1946, he had accumulated enough points to be shipped home, and by February, he was in the inactive reserves and joined the Ohio National Guard. He received his commission to second lieutenant while in the Guard, and one and a half years later he was promoted to first lieutenant.

The Ohio National Guard was activated in 1951 for the Korean War; Mr. Eversole was recalled in 1952. He was assigned to the 40th Infantry Division, 980th Field Artillery battalion, as a forward observer. They were stationed a few miles south of the 38th parallel*. Mr. Eversole said that in many ways, the Korean War was worse than World War II, mainly because of the cold. It was difficult to wage war when their machine guns and their rifles would freeze up.

After a year in Korea, he was sent to Fort Campbell, Kentucky, in 1954, where he involuntarily went through jump school. In 1956, he was back in Germany for a year. During that time he was promoted to captain. Following a reduction in forces, he was discharged. In 1958, he enlisted in the regular Army as a sergeant. In 1963, he was back in Korea for a year as an artillery sergeant. In 1965, he returned to Germany. By 1967, he was in his third war in Vietnam. Although not in combat, he was shot at a few times. He was a battalion ammunitions sergeant. While in Vietnam, he reenlisted for six more years, finally retiring in 1971.

Eugene Eversole had a storied career. He was a sergeant, a second lieutenant, a first lieutenant, a captain, and a sergeant again, finally retiring as a captain. He had served for twenty-seven years, saw a lot of the world, and earned a Bronze Star with an oak leaf, Army Commendation with one oak leaf, Combat Infantry Badge, Airborne wings, WWII Victory ribbon, European Occupation ribbon, National Defense Service Ribbon, UN Ribbon, and ribbons from the Korean and Vietnamese governments.

Lester L. Sharrit
U.S. Army, Navy, Marine Corps

Lester Sharrit joined the Army Reserves at seventeen. When World War II started a month later, he left the Army to join the Navy. He went to boot camp at Great Lakes Naval Training Center [Chicago, Illinois] and then was shipped to Hawaii. He was in aviation ordnance, cleaning, repairing, and loading aircraft weapons. He was then assigned to the USS Nassau, a converted oil tanker with a flight deck. The ship plied the Pacific, refueling ships, usually with a destroyer escort. The Nassau was a slow-moving ship, full of oil, an easy target for the Japanese. It was attacked by kamikaze planes [Japanese suicide missions], but never hit. When the war ended, Mr. Sharrit came home and joined a Marine Reserve unit.

When the Korean War began, his unit was activated and he was back to war. He sailed on a troopship to Pusan* and then to Kimpo Air Base, where he would again be in aircraft ordnance. When the Army and the Marines began the long retreat south from the Chosin Reservoir* in December, 1950, the Chinese army, which had crossed the Yalu River, into North Korea, to support their fellow Communists, continued to pursue them. The Army and the Marines would have been overpowered, said Mr. Sharrit, if the planes from his airbase—which his unit kept armed—had not constantly attacked the Chinese soldiers. Mr. Sharrit spent five years in the Marine Corps, with three or more of those years in Korea. When he returned to the States, he decided that after having been in the Army, the Navy, and the Marines—and having been in two wars—he would retire from the military. Today, he spends his free time volunteering for Ohio Reads, helping young children with reading problems.

Clem H. Whittebort
U.S. Army

Clem Whittebort earned a Purple Heart, a Bronze Star, a Pacific Theater Ribbon with Bronze Stars, an American Victory Medal, and a Good Conduct Medal in World War II. He had served in the Pacific Theater in the Fiji Islands, Guadalcanal, Bouganville, and the Philippines. So he saw plenty of action in WWII. When the atomic bombs ended the war, he came home.

When he got back to civilian life, he joined the National Guard. He had been a private first class in the Army and became a warrant officer in the Guard. His unit was activated and sent to Korea, where he was an officer in a tank battalion, attached to the 7th Division. He was not in combat in this war, but worked in personnel behind the lines. He liked the military life, he said. You get to go places and meet nice people. He said that it is up to soldiers like him to run the Army, because you couldn't leave it to the generals. When he got home, he decided it was over. He had had enough, so he left the Army. When asked how he felt about his country, he said, "I just love it."

OTHER PARTICIPANTS

Andrew L. Fisher
U.S. Army

In October, 1951, I received my draft notice and on November 4, 1951, I boarded the bus which took me to Fort Wayne, in Detroit. Hundreds of us were sworn into the U.S. Army and then taken to the Fort Street Station, where we got on a train for Camp Custer in Battle Creek, Michigan. We were put in an old World War I barracks that had no heat and no hot water.

I only had the clothes on my back, because we were told that we would be issued clothes when we got there. We didn't get clothes for days and it snowed, so my shoes and socks were wet, with no way to dry them. After that miserable introduction to the Army, I went on my second train ride to Indiantown Gap Military Reservation, near Harrisburg, Pennsylvania. It was an overnight trip, so we slept in a Pullman car. I will never forget sleeping in the comfortable upper berth, after that cold, damp barracks.

For basic training, I was assigned to Company I, 5th Infantry Division. We trained in the hills of Pennsylvania, during the winter snows. Sometimes the snow was so deep that we were unable to go to the firing range or on a march. Often after spending hours out in the cold, we would have a class indoors and I would fall asleep. I hoped that I would never be in a 30-caliber machine-gun squad, because I slept through that entire class.

We were just about through with basic when I got pneumonia. I spent a few weeks in the base hospital and didn't graduate with my class. They all went to Germany and I went to another basic group to finish my training. Four of us decided to sign up for Officers Candidate School. Two did not meet the requirements; one had signed up before, then changed his mind, so he was disqualified—that left me. I told them that I would be back, but never returned, because I learned that I would have to extend for an additional year. Some of us also went to sign up for Airborne, but I came to my senses before I got to that office.

I did sign up for Leadership School after finishing basic. We were told that if we were accepted in Leadership School, we would become corporals and if we finished in the top five of our class, we would get sergeant's stripes. But I didn't get either the corporal's stripes upon acceptance or the sergeant's stripes, when I finished in the top five of my class. There never was an explanation. After graduation, we waited a week for orders. There were one hundred of us in the class. Ninety-six got orders for the Far East command, which meant Korea. Two got orders for Alaska and two got orders for USFA, the United States Forces in Austria.

I was one of the two going to Austria. I don't know how that happened—I didn't have friends in high places, I had no special talent, I didn't speak German, and it wasn't done alphabetically, as is usual in the Army. I wondered then, and I wonder now, why I was spared the war and the terrible weather in Korea. For almost sixty years, I have wondered what happened to those ninety-six classmates. How many never lived to see twenty-one, how many were wounded, how many were missing in action or prisoners of war? Of course, there was no way to ever learn of their fates.

Not only did I get sent to Austria, the most beautiful little country in Europe, but I was sent to Camp Roeder, ten minutes from Salzburg, the most beautiful little city in Austria. Being in Austria was not like being in the Army, since we were considered to be guests and not occupation troops. We were just there to be on the other side of the Inn River from the Russians, to keep them from taking over Austria. Austria was divided among four nations, the French, British, Russians, and Americans, as was Germany. The capital city, Vienna, was in the Russian zone, but it was a four-power city, as was Berlin. There were a few thousand of us and hundreds of thousands of Russians, so we would have lasted just hours, had they decided to cross the river.

While my former classmates were fighting and dying in Korea, I was seeing Europe at Uncle Sam's expense. It seemed unfair, but as anyone who has been in the service knows, you go where you're told to go and do what you're told to do. Austria and particularly Salzburg was such good duty, that even the draftees were extending for a year to stay there.

In 2002, I had the unique opportunity to serve my country and repay the University of Toledo for the education that I received. The university and the Library of Congress are partners in the Veterans History Project, whose goal is to collect and preserve the military histories of the men and women of Northwest Ohio and Southeast Michigan. I volunteered for that effort and interviewed more than 500 veterans. After fifty years, I finally made a contribution to my country.

Clenastine J. Hamilton
WAC

Clenastine Hamilton knew what she wanted her life to be. She wanted to go to college and she wanted to be a nurse. She achieved both those goals, in spite of poverty and segregation.

I was born in Nashville, Tennessee, and went to high school there. I wanted to join the Army, but I needed my mother's approval because of my age. She finally agreed when I was eighteen. I joined the WACs, the Womens Army Corps, and was sent to Fort Lee, Virginia, where I took basic training for two months. The first night in the barracks they said we were going to have a party and we did, a GI party. That's where you scrub the floor on your hands and knees.

Was the women's training the same as the men's?
Yes, we even had to go into the building full of gas and take off our gas masks and say our name, rank, and serial number—just like the men did.

Were you ever sorry, in basic, that you had joined the Army?
Yes, I cried the day I had to clean out the mess hall grease trap.

Where did you go from Fort Lee?
I went to Fort Ord, California, for eight months and then to Fort Lewis, Washington, where I worked on troop shipments to the Far East Command, mainly Japan and Korea. Next, I went by troop ship to Yokohama, Japan.

Did you have to sleep down in the hold like the troops did?
No, we had a compartment where a few of us women slept. I went to Yokohama, where the people had seen black men but had never seen a black woman, so they would follow me around, just to look at me. We were then sent to Camp Zama in Osaka, where I outranked the men in the office.

Did the men accept the fact that you outranked them?
Yes, it was not important.

What were your duties?

I interviewed the wounded men back from Korea and prepared the papers for awarding them the Purple Heart. This experience changed my life. It was then that I decided that I would be a nurse.

Did you have a furlough to travel in Japan?

No, we would just take day trips, when we learned to use the subway. Once, we went to Tokyo, where we went shopping in the Ginza [upscale shopping district in Tokyo]. It was like New York.

How long were you in Japan?

About one year and then I returned to the States to be discharged.

Did you use the GI Bill* to go college?

Yes. It took me five years to get a BS [bachelor of science] degree. When I became an RN [registered nurse], I came to Toledo to work at Toledo Hospital for the next twenty years. The three years that I spent in the Army set the standard for the rest of my life. I was proud to serve my country; it gave me the opportunity to succeed and it taught me about life.

Lloyd Hill
U.S. Army

Chemical and biological warfare has been used for more than 3,500 years, including the twentieth century, when during World War I, the Germans spread anthrax among Allied horses, and artillery shells filled with mustard [chlorine] gas were fired at Allied trenches. The Geneva Protocol of 1925 prohibited the use of chemical and biological weapons in warfare, but did not expressly prohibit their production and storage. The United States, in accordance with the Geneva Protocol, did not use biological or chemical weapons, but did produce and stockpile such weapons.

Lloyd Hill, now of Perrysburg, Ohio, was assigned to this little-known unit of the U.S. Army. He explains the function of the CBR, the chemical, biological, radiological unit.

I enrolled at the University of Illinois, in the college of engineering, and was a mechanical engineering candidate. Since I was interested in coal and coal-fired boilers and generators, I thought I would work for a utility. I was hired as a student engineer by Union Electric Company in St. Louis [Missouri] and spent six weeks with them, when I was drafted October 18, 1951. I really wasn't too keen about going because it had interrupted my career. I made an attempt to join the Navy, because my father had been in the Navy in World War I. I attempted to become a naval candidate for OCS [Officer Candidate School] and was accepted, but the draft board decided that they should prevail and my career as a naval officer came to an abrupt halt. I was sworn into the Army.

I went first to Camp Custer in Battle Creek, Michigan, and then to basic training at Fort Knox, Kentucky. After eighteen weeks of basic training, our group of 140 awaited orders to be sent to Korea. The names of six men not going to Korea were read off and I was one of them. As an engineer, I was assigned to Army Chemical Center, Edgewood, Maryland. I had no experience in this field, except for Chemistry 101 and 201 in college. I went home on leave and then reported to Edgewood. It was a top-secret organization and required special clearance. They had checked my background and I was accepted

I was put with a group specializing in packaging. Army Chemical Center was a manufacturer for the U.S. Army's chemical, biological, and radiological warfare devices, known as CBR. This was all new to me, but I knew from the items produced there— mustard gas, nerve gas, chemicals for flame throwers, and anthrax—that it could be deleterious to one's health. Some of the chemicals were hard to handle, which was the reason that we were sent to a two-week Army packaging course at a place called the Rossford Ordnance Depot [Rossford, Ohio].

Two of us checked into the Willard Hotel in Toledo, and then went to the Ordnance Depot, a bustling, thriving place at that time. We walked into the building and were greeted by a young lady receptionist named Lois Ann Heckman, from Bradner, Ohio. I immediately knew she was the one and was going to become my wife. We married in November, 1953, after my discharge.

After our two-week course at Rossford, we were sent back to Edgewood, where I was involved in shipping flamethrower propellant. We then spent a lot of time traveling around the country on quality control inspection tours of gas-mask component manufacturers, of which there were ten or twelve, all making different parts of gas masks. It seemed strange to me that we as Army privates were involved in high level meetings with executives of these corporations.

Were any of these plants that you visited making CBR weapons?
Not to my knowledge, but some facilities were top secret and not open to inspection by us.

According to the Geneva Protocol, chemical, biological and radiological weapons could not be used by signatory nations, but could be produced and stored. Was Edgewood actively producing CBR weapons and do you believe that they would be used if any enemy used such weapons first?
Yes, we were producing CBR weapons and I can only assume that they would be used in retaliation of an enemy first strike.

What was your unit mainly packaging?
Propellants for flame throwers.

Where were these CBR weapons stored?
They were stored in three places around the U.S.—at Edgewood in Maryland, at Redstone Arsenal in Alabama, and at Dugway Proving Grounds in Utah.

These chemicals are highly toxic and life-threatening in the event of an accident. Was there ever an accident that endangered the local population?
Yes. Dugway in Utah is far removed from any population center, but toxic gas did escape one time and killed a number of sheep. Many of these chemicals are stored in fifty-five-gallon drums, which are subject to deterioration and have been [deposited] in remote areas of the Pacific Ocean.

Mr. Hill served in a top secret unit that was largely unknown to the rest of the armed services and to the American people. For most Americans, this is probably the first time they will have ever heard of a CBR unit. Knowing that the potential for releasing highly toxic chemicals was ever-present in Korea and in succeeding wars, is frightening.

Raymond L. Ott
U.S. Navy

There were those American GIs who after serving their country,
came home, took off their uniforms, and resumed their civilian
lives, without telling their families anything about their war-
time experiences. When they passed on, their families were left to
wonder. That might have been acceptable in the past, but today
widows, children, and grandchildren want to know what their
loved ones did in the service of their country.

If the family is fortunate, the veteran had saved some of his official documents. But, as is often the case, the family members, not having been in the service, are unable to fathom what the documents are or where to look for them.

Raymond L. Ott died without ever discussing his military service with his family. Now many years later, his family wants to know about his time in the service, for themselves and for future generations. And so, with Carole Colboth, Mr. Ott's widow, we began to piece together the evidence. It is instructive to explain that process.

Mrs. Colboth had official papers in official envelopes, some of which were important to our search. The most important paper found was a DD214, Report of Separation From the Armed Forces of the United States. It contains information from time of entry into the service until discharge.

Mrs. Colboth knew her husband had been in the Navy during the Korean War, but not much more. Here is what we learned, beginning with his entry into the service.

Raymond L Ott was born and raised in Richfield Center, Ohio, a farming community, west of Toledo. He enlisted in the Navy on December 28, 1950. After boot camp at Great Lakes Naval Training Center [Chicago], he was assigned to the USS Catamount LSD 17, at its port in Sasebo, Japan.

Seaman Ott's service record indicates that he served three years, ten months, and five days in the Navy, of which three years, seven months, and seven days were at sea or in foreign ports. It also shows that he had among other medals, a Korean Service medal with three stars, indicating that he had been in four major wartime actions, none of which were named on his service record.

Further investigation revealed facts about the USS Catamount. LSD is an acronym for Landing Ship Dock. Its function was to carry troops and landing craft. From a history of the Catamount, we learned that prior to Seaman Ott's joining the ship, it took part in the September, 1950, invasion of Inchon*, the second invasion at Wonsan* in October, 1950. In November, 1950, the Catamount was sweeping mines in the harbor at Pyonyang, North Korea's capital city. In December, 1950, the Catamount took part in the evacuation of Marines from Hungnam.

After Seaman Ott joined the Catamount, the LSD took part in the following major campaigns: Communist China Spring Offensive; Second Korean Winter; Korean Defense Summer-Fall 1952, and the Third Korean Winter.

We spoke to Mrs. Colboth about her late husband.

Did your husband ever say anything about being in the Navy?
He had some former shipmates in Lorain, Ohio, with whom he would discuss the war, but never with his family.

Did Mr. Ott join any of the veterans organizations?
After some years he joined the VFW [Veterans of Foreign Wars].

It was often the case that former GIs felt more at ease discussing the military with their peers at the VFW or other veterans' organizations. But those conversations are lost forever. Although your husband saw plenty of action in the Korean War, perhaps, like other veterans, he was loath to discuss the war with his family.
He was proud of having served his country and every year he marched in the local Veterans Day parade, carrying the flag at the head of the parade. When he died, his post paid for his grave marker. That was a great honor for him and for me. And now, we make certain that there is always an American flag at his gravesite.

It is the objective of the Veterans History Project to preserve the histories of those who did the real work. We can read all about MacArthur, Truman and Eisenhower, but what about Seaman Raymond L Ott?
I was proud that Ray had been in the service and that I was the wife of a veteran. I think that veterans are special people.

George M. Rinkowski
U.S. Army & U.S. Air Force

Father George Rinkowski served his God and his country for sixty-five years. He was first an Army chaplain in the Pacific Theater during World War II and then returned to serve in the Korean War as an Air Force chaplain. After leaving the armed forces, he served as a parish priest and as the state and national chaplain for the Veterans of Foreign Wars.

Father Rinkowski was one of the eighty WWII veterans featured in my earlier book, What A Time It Was. That interview covered his service in WWII. Here we will briefly discuss his WWII service, his time in Japan, after WWII. We will then consider his time in the Korean War and his postwar work with the VFW.

Father Rinkowski was ordained in 1941. In 1945, as an Army chaplain, he was sent to the Philippine Islands, while the fighting was still raging. After the Japanese surrender, he was stationed in Japan as a traveling chaplain, where he served until 1948.

When I returned from Japan to the U.S., in 1948, I was stationed at Eglin Air Force Base, Illinois, where I was the base chaplain. During the Korean War, I was sent to Korea. I flew on a C54 [Douglas C54 Skymaster], with forty-six other chaplains. We were about 150 miles out from California, when one of the four engines quit. We continued on to Hawaii on three engines. From Hawaii, we continued on to Korea.

Were you there when the fighting was going on?
Yes, they were moving up the peninsula all the way to the border of China.

Did you follow the troops all the way north?
No, I was with the Air Force, so I stayed on the base.

What does a military chaplain do?
He does everything that a parish priest does and also tends to the casualties. See that Mass kit on the table? I carried that with me wherever I went.

When did you leave Korea?
I returned to the States in 1953.

What was your next assignment?

I was sent to Chanute [Air Force Base], Illinois, where I was the base chaplain. I was a major, so wherever I was stationed, I was the head chaplain.

Was Chanute Field your last assignment?

Yes, I left the service in 1956.

You were forty-two years old with about eleven years of service. Did you consider staying in the military?

No, I wanted to return to being a parish priest, so I came back to Toledo, where I was assigned to a parish. One of my assignments was at St. Hedwig Church, where I stayed for seventeen years.

But you continued to serve as a chaplain at the VFW [Veterans of Foreign Wars] and the DAV [Disabled American Veterans].

I was the chaplain at one time for five different veterans' organizations. I was also honored to be the State and National Chaplain for the VFW.

Father Rinkowski retired at the age of seventy-three from the full-time priesthood. He was eighty-nine when he was interviewed by the Veterans History Project, and he died at age ninety-three. Requiescat in Pace.

Yunhoon Chung

Mr. Chung's life story is the stuff of a Hollywood movie. He was one of the millions of innocent Korean civilians whose lives were uprooted by the Korean War. He has lived in Manchuria, North Korea, South Korea, and the United States. In that time, he has lived under the Russians, Japanese, Nationalist Chinese, and the Communist Chinese. He was a fugitive hiding from the North Koreans and finally he made his way to the United States, where he went to college, became an inventor, an industrialist, and a restaurateur.

Mr. Chung's family had lived in North Korea for 300 years until they moved to Manchuria, where his father owned a textile business, and where Mr. Chung was born in 1930. His father also started a school and was its teacher. Since Manchuria was then under Japanese rule, Japanese was the language taught in school and the family name Chung had to be changed to Yoshitake.

In 1945, at the end of World War II, the Japanese left Manchuria and the Russians occupied the country. They were followed briefly by the Nationalist Chinese. In 1946, the Chinese Communists took over China and Manchuria. In 1947, the Chung family left Manchuria and moved to back to their homeland, Korea, where the Russians occupied the northern part and the United States the southern part. Following the Russian refusal to allow free elections in their zone, the country was divided into North and South Korea. North Korea was cut off from the rest of the world and Mr. Chung, then in college, was stuck in North Korea.

In 1949, Mr. Chung escaped into South Korea, but his family did not. He saw his father once in 1951, but he never again saw or heard from his mother, his sisters, or any other relatives in North Korea. People who defected from North to South Korea were subjected to intense interrogation and even torture to determine if they were communist spies. He was undergoing interrogation when it was learned that he spoke English. He was taken to the CIA office at Kimpo Air Base, where he was offered a job. He wanted to go to college, so he turned down the offer and worked instead as a laborer while going to school. The Americans later offered to train him for a job as a weatherman, which he accepted, but when the war began in June, 1950, that program ended.

Mr. Chung was living in Seoul when the war began. When the North Koreans captured Seoul, he had to go into hiding. After a few days, he was caught when he came out to find food. Three times he was caught by the Communists, who conscripted him into their army and three times he escaped.

Mr. Chung continued to live in Seoul, after the North Koreans were driven back north, but in January, 1951, they again captured Seoul. Mr. Chung made his way south to Taegau [now called Daegu], where he got a job with the British as a court-martial court interpreter. In January, 1952, he began to make a serious effort to get to the United States. He needed a letter of recommendation, which he got from the president of the college in Hungnam, North Korea; he had to have an American sponsor, and he had to be accepted by a U.S. college. He wrote to five colleges and was accepted by the University of Michigan. In 1953, he went to the U.S. and in 1954, his future wife joined him.

Mr. Chung graduated with a mechanical engineering degree, which gave him the background for his inventions. In 1960, he invented a machine to package sugar used in restaurants. He eventually built International Automated Machines, Inc., a 110,000-square-foot plant in Perrysburg, Ohio, and had 200 employees.

Mr. Chung commented on the wisdom of General MacArthur's plan to invade China. He said that Koreans generally feel that President Truman was right to prevent MacArthur from invading China and was right to fire him, in spite of the fact that MacArthur was looked upon as a god by Koreans.

When asked if North and South Korea would ever unite, he promptly answered yes. He said that the North Korean pool of low-cost labor and the South Korean's industrial know-how would make the united country an economic power to rival Japan.

Mr. Chung represents the typical Korean who was caught up in the war that raged up and down the peninsula. Today he is an industrialist of some means, but when he was a poor young man, in Korea, with no money, no family and on the run, he was like the millions of other innocent civilians struggling to stay alive in a war that ravaged their country.

Dr. Su-Pa Kang

*"Not again, now we are fighting our own people,"
said Su-Pa Kang when North Korean forces
crossed the border between North and South Korea
on June 25, 1950, intent on conquering South
Korea and uniting the two under communist rule.
World War II had been over fewer than five years
and the defeat of Japan finally ended their thirty-
five-year annexation of Korea. The South Koreans
were just beginning to recover when yet another
war was visited upon their still-poor country.*

*Su-Pa Kang was a boy of ten when the Korean
War began. He lived with his family in the city of Gwangju, in the far southwest corner of
South Korea. Gwangju was an industrial city of 300,000 at the time, with plants producing
textiles and batteries. His was an upper middle-class family, but his father was in the
underground and spent nine years in jail for his activism against Japanese rule.*

*When World War II ended and Korea was freed, his father felt that after years of
foreign domination, the people were not yet ready for democracy, but should transition
from a strong leader to democracy.*

When the North Korean forces drove south all the way to the Pusan perimeter,
they occupied the city of Gwangju. Because Su-Pa Kang's father was Vice-Governor of
Chollanamdo Province, and his family was politically influential, they had to go into
hiding. They were helped by a North Korean officer who had once been a worker on
his father's farm before he joined the Communists. Two of Su-Pa Kang's uncles were
not so fortunate; they were taken to North Korea and were never seen again. When the
United Nations forces landed at Inchon* and cut off the North Koreans, in and around
Gwangju, those who were trapped became guerillas, often hiding in the mountains
by day and coming down at night to steal food. They were eventually burned out of
the mountains.*

What is the relationship of the Koreans and the Chinese?
After centuries of being neighbors, politics aside, the Korean people have an affinity
for the Chinese people but not the Japanese.

What was the relationship of the Korean people and the United Nations forces?
The Korean people welcomed the UN forces.

What are the prospects of a united Korea?

It would take a revolution in North Korea, and South Korea would initially suffer as a result, much like when West Germany, which was prosperous until it united with the poor East Germans. North Korea is interested only in military spending, while the people are starving.

When East and West Germany were first divided [after WWII], people crossed over to the West by the tens of thousands. Did that happen or does that now happen in Korea?

The North Korean people cannot cross into South Korea because of the Demilitarized Zone and the North Korean people have absolutely no contact with the outside world at all.

When and why did you leave Korea?

I had graduated from the Chonnam National University's medical school in [Gwangju] South Korea and came to the States for further post-graduate medical training in 1968 at age twenty-eight. I interned at [the former] Mercy Hospital [Toledo], was a resident at [the former Washington] D.C. General Hospital and the Medical College of Ohio and had a fellowship at the Cleveland Clinic.

Do you still have relatives in South Korea and do you visit them?

I have five sisters and their families and many cousins there and I visit them every few years.

How has South Korea prospered since the war?

My hometown of Gwangju has grown from 300,000 to [around] one and a half million. Seoul, the capital, now has a population of eleven million. South Korea has prospered because of its political system and because the people are diligent. South Koreans are hard-working people who are doing everything possible to educate their children.

APPENDICES

TERMS USED

To avoid re-defining the terms, battles, planes, and other references in *Thirty Below on Christmas Eve*, such words are starred with the definitions given here. A brief history of the various personalities who played important roles in the Korean War is also provided on page 131.

38th parallel: circle of latitude, 38 degrees north of the equator, passing through Europe, Mediterranean Sea, Asia, Pacific Ocean, North America, and the Atlantic Ocean. After World War II, the occupation of the Korean peninsula, which had been under Japanese control since 1911, was divided between the United States and the Union of Soviet Socialist Republics. The 38th parallel was made the dividing point as it bisects the peninsula. On June 25, 1950, North Korea troops crossed the parallel, invading South Korea and starting the Korean War. When the June, 1953 armistice ended the conflict, the 38th parallel became the dividing line/border, permanently separating the two countries.

40&8 Society: officially "La Société des Quarante Hommes et Huit Chevaux," or the Society of Forty Men and Eight Horses. It takes its name from the box cars used to transport men to the French front during World War I. The cars could hold forty men or eight horses.

Amtrack: flat-bottomed military vehicle that moves on tracks on land or water.

B26: twin-engine, light-attack bomber built by the Douglas Aircraft Co.; later known as the Douglas A26 Invader.

BAR: Browning automatic rifle. Originally used in World War I, the BAR is often defined as a light machine gun.

Bloody Ridge: battle fought from August 18-September 5, 1951 (during the Stalemate Period), north of the 38th parallel, in the Korean highlands, for control of a ridge believed to be used to call in artillery fire on a United Nations supply road. After ten days of back-and-forth assaults, the North Korean forces withdrew and set up some 1,500 yards away on a hill that would be known as Heartbreak Ridge. Some 2,700 United Nations troops and perhaps as many as 15,000 North Koreans were killed in the battle.

Bug out: military slang for leaving in a hurry or retreating during a military action.

Ch'ongch'on River: battle in North Korea, some fifty miles south of the Chinese-Korean border, from November 25-December 2, 1950, during the coldest North Korean winter in one hundred years. Successful surprise attacks at Onjong and Unsan enabled the Chinese to defeat the South Koreans and the U.S. 1st Cavalry Division, along with destroying the right flank of the U.S. 8th Army, forcing the troops to retreat to the Ch'ongch'on River area. Then the Chinese withdrew. UN tacticians and General MacArthur were convinced that only some 30,000 Chinese troops remained in Korea; bombing along the Yalu River to destroy bridges would ensure that they could not be reinforced. But the Chinese 13th Army, some 230,000 strong, was in place, along with some 150,000 additional Chinese troops heading to the Chosin Reservoir.

The PVA 13th Army launched a surprise attack on the U.S. 8th Army; at the battle's end, the U.S. 2nd Infantry Division, the Turkish Brigade (5,400-man unit attached to the U.S. 25th Infantry Division) and the South Korean II Brigade were out of action and the 8th Army was reduced to two corps. The 8th Army's General Walton Walker ordered his troops to retreat; they were joined by the Marines from the Chosin Reservoir battle in the longest retreat in U.S. history. Walker died on December 23 and was replaced by General Matthew Ridgeway. The Ch'ongch'on River battle would be China's best success, but the 13th Army troops, undersupplied and almost starving, were not able to mount further offensive operations.

Chosin Reservoir: November-December, 1950 battle site, in northeast North Korea. Chinese troops entered North Korea and engaged the UN troops, including the 1st Marine Division, the 3rd and 7th Infantry Divisions, and a small contingent of (British) Royal Marine Commandos. The reservoir region lies about 3,200 feet above sea level with temperatures of -35 degrees in the winter. On November 27, the Chinese attacked, catching the UN troops by surprise and cutting them off. China sent additional troops, and by December 1, UN forces were in retreat, in a massive evacuation ending at the port of Hungnam, on the North Korean east coast. Survivors of the 30,000 UN troops are sometimes called the "Chosin Few" or the "Frozen Chosin."

Combat Infantry Badge: given to an Army infantryman who has satisfactorily performed infantry duties. The soldier must be assigned to an infantry unit when the unit is engaged in active ground combat, and must actively participate in ground combat.

Corsair: F4U Corsair, carrier-launched fighter plane, built by Chance Vought.

DMZ: Demilitarized Zone, 2.5-mile wide strip of land on either side of the 38th parallel, forming the boundary and separation of North and South Korea. The 160-mile-long DMZ is the most heavily militarized border in the world.

F80/P80: F80/P80 Shooting Star, built by Lockheed Corp., the first jet aircraft used by the U.S. Army Air Force. The jets were used against the MiG-15 in Korea but were not as effective and were replaced by the F86 Sabre.

F84/F86: F84 Thunderjet, turbojet fighter aircraft built by Republic Aviation. The F84G was the Air Force's main strike plane during the Korean War. The F86 Sabre, made by North American Aviation, was a transonic jet that encountered the Soviet-made MiG15 over the Korean skies.

FECOM: Far East Command. U.S. Army headquarters in Japan.

Geneva Conventions: four treaties and three protocols that establish standards for the humane treatment of war prisoners, including the banning of torture, and of victims of war. The first treaty was adopted by twelve countries in 1864; the second in 1906; the third in 1929, and the fourth in 1949. Two protocols, in 1977 and a third in 2005, were designed to cover civilians and objects.

GI Bill of Rights: passed in 1944, the GI Bill provided college and vocational assistance to returning World War II veterans, as well as loans to purchase homes or start businesses. The bill has been amended and expanded many times in the following years.

Heartbreak Ridge: battle fought ten days after the battle of Bloody Ridge ended, from September 13-October 15, 1951, during the stalemate period. The North Koreans had retreated just 1,500 yards from Bloody Ridge to a ridge seven miles long, a few miles north of the 38th parallel. The 2nd Infantry Division's acting commander, Brigadier General Thomas De Shazo, and Major General Clovis E. Byers, X Corps commander, underestimated the strength of the North Koreans and sent only the 23rd Infantry Regiment and an attached French battalion straight up the ridge. The troops would bombard the targets, then climb up the rocky terrain to take out bunkers. The North Koreans would counterattack in waves, forcing the UN troops to retreat back down the ridge.

On September 27, the 2nd Infantry's new commander, Major General Robert N. Young, halted the futile assaults and called in the division's 72nd Tank Battalion to cut the North Koreans off from further reinforcements. The armored attacks turned the tide for the UN troops and the French captured the last North Korean position on October 13. Over 3,700 U.S. and French troops were killed, along with an estimated 25,000 North Koreans.

Inchon: September 15-19, 1950 battle site, on the Korean western coastline, fewer than forty miles west of Seoul. The amphibious invasion involved some 75,000 troops and 261 ships. Inchon was a decisive victory for the UN forces, the majority of which were U.S. Marines under the command of General Douglas MacArthur. Inchon ended the string of North Korean victories and enabled UN forces to recapture Seoul, the South Korean capital. Now spelled Incheon.

LST: landing ship/tank. Ships that supported amphibious military operations.

MASH unit: Mobile Army Surgical Unit, a fully functioning hospital on the battlefield. First established in 1945, MASH units were used during the Korean War, and proved to be quite successful at treating the injured more quickly. The term is known today because of M*A*S*H, first a novel by Richard Hooker, then a feature film, and finally the television series which ran eleven years, eight years longer than the war itself. The final show, "Goodbye, Farewell and Amen," held the most-watched television show record for twenty-seven years, losing the title in 2010 to the Super Bowl. The M*A*S*H finale, however, still holds the ratings and audience-share records.

MiG-15: Mikoyan-Gurevich MiG-15, a swept-wing jet fighter, built in the Union of Soviet Socialist Republics and flown in the Korean War by Soviet pilots (with planes disguised as North Korean or Chinese aircraft) as well as the Chinese military.

MLR: main line of resistance, military term for the major defensive position of opposing armies.

Old Baldy: Hill 266, one of a series of hills along a ridge north of the 38th parallel along the MLR, central to a series of five battles over ten months, starting in June, 1952, to capture and hold twelve outposts along the ridge. Hill 255 would become known as Porkchop Hill. The final Old Baldy battle, March 23-26, 1953, included troops from Colombia, the only South American country to contribute soldiers to the UN forces.

Operation Little Switch: exchange of sick and wounded prisoners in April-May, 1953. Some 6,600 Chinese and North Korean prisoners were released by the UN in return for 684 coalition prisoners, including 149 Americans.

P51: P51 Mustang, a long-range fighter aircraft built by North American Aviation and used during World War II and the Korean War.

Panmunjon: former village on the border between North and South Korea, on the northern side of the military demarcation line, which runs through the Demilitarized Zone (DMZ). United Nations forces met with North Korean and Chinese officials at Panmunjon from 1951-53 for truce talks. The final agreement was signed on July 27, 1953. The building where the armistice was signed still stands, although a new site, the Joint Security Area, was built one kilometer from the village, which eventually disappeared.

Pork Chop Hill: battles fought from March-July, 1953, for one of a series of hills along a ridge north of the 38th parallel along the MLR, during the period when the cease fire and armistice were being negotiated. The first battle, April 16-18, 1953, saw the 7th Infantry Division, along with South Korean, Ethiopian, and Colombian troops, finally secure the hilltop, clearing the Chinese from the trenches and bunkers. The second battle began July 8, after the 7th Division had rebuilt its defenses and while final details on the armistice were being negotiated. Both sides attacked and counter-attacked until July 10. The next morning, the hill was abandoned to the Chinese and the 7th Infantry withdrew under fire. The armistice was signed on July 27.

Pork Chop Hill was a costly battle as four of the thirteen U.S. company commanders were killed, along with almost 250 U.S. troops.

POW: prisoner of war. Although both the Communist and UN forces were committed to the Geneva Conventions, POW treatment varied greatly. The North Koreans saw South Korean POWs not as enemies but fellow Koreans who had been misled into fighting and were therefore not covered by the Conventions. Many North Korean POWs claimed to have been forced to fight and did not want to be repatriated to the North at the war's end. Likewise, many of captured Chinese soldiers had been Chinese Nationalists and did not want to return to Communist China.

The exact number of South Koreans held in the North both before and after the war remains unknown. Veterans Administration documents put the number of American POWs at 7,140. Of those, 2,701 died in captivity for a 38 percent mortality rate. Following the July, 1953 armistice, the UN command repatriated 75,823 Communist POWS while 12,773 UN POWs were returned. Only 10 percent of the South Koreans missing in action were returned.

Twenty-one Americans refused repatriation and stayed in North Korea, although the New York Times reported in 1996 that Department of Defense declassified documents stated that as of the end of 1953, more than 900 American troops were alive but never released by the North. The Pentagon did not confirm the Times report, saying it had no clear evidence that Americans were being held against their will.

The South Koreans claim 387,744 of their citizens were either abducted or remain missing.

Punchbowl: deep circular valley in the North Korean Taebeak mountains. Site of fierce fighting between U.S. Marines and the Chinese from June through the end September, 1951, during the Stalemate Period.

Pusan: city on the southeastern tip of South Korea, one of only two cities not captured by the North in the first months of the war. UN forces set up a 140-mile area, the Pusan Perimeter, where U.S. forces, joined by South Korean and British army troops, held out against the North Korean advance from August 4-September 15, 1950. After the UN amphibious attack at Inchon, the North Korean army collapsed and retreated. Now spelled Busan, Pusan was the temporary capital of South Korea.

R&R: Military slang for "rest and recuperation" or "rest and relaxation," planned leave time.

Shoepacs: winter footgear first provided in WWII. The boot had a heavy rubber bottom, leather upper, and felt innersole or liner, with waterproof seaming. Because of the rubber bottom, soldiers had to change socks frequently to prevent moisture from freezing inside the boots. In Korea, however, the rubber-bottomed shoepacs were woefully inadequate for the extreme winter weather, despite the addition of two pairs of insoles and six pairs of wool socks. The soldiers were told to change socks frequently, a difficult task in below-zero weather or when changing battle positions.

Before the winter of 1951-52, the shoepac was replaced by the Mickey Mouse boot, which had two layers of rubber with a layer of wool pile insulating material in between. This layering meant both air and moisture were kept out of the boot, preventing heat from escaping the foot, protecting the wearer from frostbite down to minus-twenty degrees. The nickname came from the boots' color and shape.

Stalemate period: from July 1951 to July 1953, during which the UN and Chinese forces fought but exchanged almost no territory. Truce or peace talks continued throughout the period, with the Chinese and North Koreans testing the UN resolve to remain in the fight with military and psychological operations.

Several major battles were fought during the stalemate, including Heartbreak Ridge, Old Baldy, White Horse, Triangle Hill, Hill Eerie, the Hook, and Porkchop Hill, along with the sieges of Outpost Harry.

Taejon: Now spelled Daejeon, a major South Korean city and site of one of the first battles, July 14-21, 1950, where US Army forces fought the North Koreans who were advancing quickly down the Korean peninsula. After a three-day struggle, the Americans retreated, but by delaying the North's further advance, UN forces were able to establish and hold the defensive perimeter around Pusan, finally stopping the Northern advance.

Major General William F. Dean, commanding the 24th Infantry Division, was separated from his men and taken prisoner by the North Koreans. He remained a POW until the end of the war, after which he was awarded the Medal of Honor.

Truman Year: a wartime extension of an additional year of service beyond a service member's original discharge date.

Wonsan: port and naval base in eastern North Korea. The city was under continuous siege and bombardment from March, 1951 to July, 1953, and was largely destroyed at the war's end.

Yalu River: the 491-mile-long river that forms over half of the border between North Korea and China.

— Molly Schiever

PERSONALITIES

Kim Il-sung
President of the Democratic People's Republic of Korea

The Democratic People's Republic of Korea was founded in North Korea in September, 1948, with Kim Il-sung as prime minister. Kim remained an absolute dictator until his death in 1994, at eighty-two years old.

When Kim Song-ju was born in 1912, Korea was ruled by Japan, and his family lived barely above the poverty level. At seventeen, he became a member of an underground Marxist group and was subsequently jailed for his anti-Japanese activities. He joined the Communist Party of China and a small guerilla group, which he commanded at age twenty-four. In 1935, he changed his name to Kim Il-sung, which means, "become the sun." In 1940, to avoid capture by the Japanese, he escaped into the Union of Soviet Socialist Republics. He was trained by the Russian Communists and became a captain in the Russian army. Many of Kim's early exploits, however, used to develop and embellish his cult of personality, are difficult to confirm with accuracy.

At the end of World War II, the Russians occupied what is now North Korea and set up a puppet government, naming Kim head of the People's Provisional Committee. He founded the Korean People's Army, equipped by the Russians. North Korean pilots were sent to China and Russia to train in the new MiG-15* jet planes.

The Democratic People's Republic of Korea was established in September, 1948, four months after the Republic of Korea had been created in the south. In October, the Soviets declared the DPRK the only legitimate government for the entire Korean peninsula. A year later, North Korea had become a full-fledged communist dictatorship, with Kim now calling himself the "Great Leader."

The North Koreans crossed the 38th parallel* on June 25, 1950, driving deep into South Korea before United States troops, under the mandate of the United Nations, entered the war in July. Kim led North Korea through the war, but even with the help of the Chinese army and Russian air power, his forces failed to conquer South Korea. The two Koreas remain in a state of suspended hostilities today. During the later years of Kim's reign, the North suffered numerous economic crises as the country fell into even deeper repression.

In 1994, after beginning North Korea's nuclear development program, Kim Il-sung died following a massive heart attack. At eighty-two, he'd outlived Russia's Stalin by forty years, China's Mao by almost twenty, and remained Great Leader through the terms of six South Korean presidents, nine U.S. presidents, and twenty-one Japanese prime ministers. He was succeeded by his son, Kim Jong-il, who was then fifty-three years old.

With the death of Kim Jong-il in December, 2011, the ruling North Korean Workers' Party issued a statement suggesting, as the New York Times noted, that his son, Kim Jong-un, was to be the apparent successor.

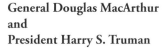

General Douglas MacArthur
and
President Harry S. Truman

General Douglas MacArthur was one of the most famous military men in American history. He had graduated first in his class at the United States Military Academy in 1903, then served in the Philippines before World War I, where he had received America's highest military honor, the Medal of Honor. He served with distinction in World War I in France and retired from the Army in 1937. He was recalled to active service in July, 1941, and named commander of the U.S. Army Forces in the Far East, but following the invasion of the Philippines by the Japanese, his forces were to retreat to Bataan and eventually Australia. There he was named Supreme Commander, Southwest Pacific Area, and directed his troops from island victory to island victory. When the war ended in the Pacific, he officially took the surrender of the Japanese on the battleship Missouri.

MacArthur was then sent to Tokyo, where he oversaw the occupation of Japan from 1945-51, supervising the rebuilding of Japan's war-damaged country, and helped establish a democratic government in the nation previously ruled by an all-powerful emperor. When the North Koreans invaded on June 25, 1950, he was appointed the commander of the United Nations forces. At the war's outset, the UN-sanctioned troops were driven all the way down the peninsula to Pusan* but by August, UN forces greatly outnumbered North Korean men and arms. MacArthur made the risky decision to send UN forces by sea to a successful landing at Inchon*. On October 1, South Korean forces had crossed the 38th parallel* into North Korea while MacArthur planned another amphibious landing in the North, despite a previous UN Security Council top-secret memorandum forbidding any such action if no Chinese or Russian forces had entered North Korea or had threatened to do so. On October 25, the Chinese warned that further advancement towards the Yalu River border with China would be seen as a military threat.

MacArthur had met with President Harry Truman on October 15, dismissed the Chinese threat, and said instead he hoped to be able to withdraw American forces by the beginning of 1951. Although the Chinese had already begun to cross the river, MacArthur, after surveying the 8th Army's front lines on November 24, saw no evidence of a Chinese buildup. The next day, the Chinese and North Koreans attacked and overran the South Korean troops and the U.S. 2nd Infantry Division at the Battle of the Ch'ongch'on River,* forcing the 8th Army to begin the longest retreat in U.S. history. On November 27, the Chinese army attacked the 1st Marine Division and a 7th Infantry Division regimental combat team at the Chosin Reservoir,* again driving the UN forces into retreat. Despite MacArthur's decision not to order an attack or withdrawal prior to the devastating Chinese attacks, he was awarded the Distinguished Flying Cross and honorary combat pilot's wings.

By March, 1951, the 8th Army had recaptured Seoul, the South Korean capital, and military leaders were debating invading the North or pressing for peace. MacArthur issued his own statement, calling on the Chinese to admit they had been defeated. Eleven days later, a Republican representative read aloud on the House floor a letter from MacArthur highly critical of Truman's limited-war strategy. The following day, Truman called together the secretaries of State and Defense, along with the Joint Chiefs of Staff to discuss what to do about MacArthur's actions. On April 8, the joint chiefs did not agree to MacArthur's removal from command but felt it was correct "from a purely military point of view." The general's fate was now in the president's hands.

President Truman was not nearly as famous as MacArthur. He had been an artillery captain of no particular repute in World War I, then a county commissioner, a U.S. senator, and in 1944, he became President Franklin Roosevelt's running mate. Following FDR's death in April, 1945, Truman ordered the use of atomic bombs to end the war with Japan. He was credited with adopting a policy of containment, called the Truman Doctrine, designed to halt the surging communist movements in Europe and Asia. Truman's Marshall Plan provided monies to help rebuild war-ravaged western Europe, while the Berlin airlift took fleeing Germans out of the communist East and brought in supplies to the isolated city.

When the communist North invaded South Korea, Truman's policy of containment applied to the defense of South Korea and was made clear in his directive to General MacArthur. Initially MacArthur believed the Chinese wouldn't attack, which had led to the huge losses in November and December, 1950. The idea of a limited war was not acceptable to MacArthur who wanted to carry the war across the Yalu River and to engage China with ground troops and air strikes, despite the fact that his official orders were not to cross the 38th parallel. He'd also argued for the use of nuclear weapons against the Chinese, and for his right to make decisions on the direction of the war. His letter criticizing the Commander in Chief was a direct challenge to presidential authority.

Harry Truman had shown his own mettle when he took the responsibility for the atom bombs and when he challenged the Russian Communists with the Truman Doctrine, the Marshall Plan, and the Berlin Airlift. Now he had to make the unpopular decision to fire one of the most famous military men in U.S. history for insubordination. On April 11, 1951, General MacArthur was relieved of his command, replaced by General Matthew Ridgeway.

MacArthur came home to a ticker tape parade and a speech on April 19 before an a joint session of Congress where he concluded with the words, "Old soldiers never die, they just fade away." He was hailed as a potential presidential candidate on the 1952 Republican ticket, but congressional hearings in May and June that determined he had defied the orders of the president and thus violated the Constitution cooled political enthusiasm. MacArthur endorsed Ohio's Robert Taft instead. The party, however, picked another famous general, and Dwight D. Eisenhower went on to win the presidency. MacArthur died on April 5, 1964, at age eighty-four, and was given a full state funeral, including lying in state in the U.S. Capitol.

Truman survived his most unpopular decision, which moved Senator Taft to call for his impeachment, but the Korean stalemate was also contributing factor to his decision not to run for reelection in 1952. He returned home to Independence, Missouri, turning down all financial opportunities or commercial endorsements. He lived on his army pension of $112.56 a month until earnings from several land sales helped supplement what remained after taxes on the profits of his memoirs. In 1958, Congress passed the Former Presidents Act, providing a lifetime pension of $25,000 for Truman and former president Herbert Hoover. On December 26, 1972, Truman died at age eighty-eight, and despite leaving office with one of the lowest approval ratings ever, is now routinely listed among the top ten American presidents.

Mao Zedong and the Korean War

Two days after the North Koreans invaded South Korea, President Harry Truman sent the Seventh Fleet to the straits of Taiwan, to protect the Nationalist Republic of China. In August, 1950, Chinese Premier Zhou Enlai informed the United Nations that China was concerned about the impact of the Korean question on Chinese security. Truman dismissed the warning as blackmail.

On October 8, 1950, Mao Zedong, Chairman of the People's Republic of China, gave the order for Chinese soldiers to cross the Yalu River border with North Korea to join the army of the Democratic People's Republic of Korea. The United Nations forces were on the move north, but during a meeting between Truman and General Douglas MacArthur, the general said there was little risk of Chinese intervention and that their opportunity to aid the North Koreans had passed. By the time the final cease-fire was signed in July, 1953, over a million Chinese had fought alongside the North, with up to 400,000 killed, more than North Korean losses.

Mao's oldest son, Mao Anying, (pictured with his father, above) volunteered to serve in the Chinese army, although Chinese leaders were reluctant to let him fight in Korea. His father did not intervene and Mao Anying crossed the Yalu River on October 25, 1950. Stationed in caves, his unit was well protected from UN airstrikes. But on November 25, South African planes dropped napalm bombs, killing the twenty-eight-year-old Anying and another officer who were cooking lunch in violation of Chinese army regulations. Mao was not told of his son's death until January, 1951, and General Peng Dehuai fell from favor permanently.

Lewis Burwell Puller
U.S. Marine Corps

He was known far and wide as Chesty, the most famous Marine in the history of the Corps. He was Lieutenant General Lewis Burwell "Chesty" Puller, five-time recipient of the Navy Cross and a legendary leader of men for thirty-seven years.

Chesty Puller was born in 1898 in West Point, Virginia. He attended Virginia Military Institute for one year, when he quit to ". . . see where the guns are." He enlisted in the Marine Corps in July, 1918, to get into World War I. He didn't get there, but became a reserve second lieutenant in 1919 and was immediately placed on inactive duty. So he rejoined as an enlisted man and served in Haiti. He spent five years in Haiti, fighting rebels and returned to the States in 1924 to be commissioned as a second lieutenant. He spent two years in Norfolk and Quantico, Virginia. In 1926, he was stationed at Pearl Harbor. It was in the 1930 American intervention in Nicaragua that he earned his first Navy Cross for heroism. Then it was back to the U.S. and back to Nicaragua, where he earned his second Navy Cross in 1932. In 1933, he did a tour in China, where he was the commandant of the Horse Marines, a unit famous since the Barbary Wars of 1805 and action in Nicaragua.

Chesty Puller did two sea duty tours, was an instructor, served in Shanghai in 1940, and commanded the 1st Battalion, 7th Regiment, 1st division at Camp Lejeune, North Carolina, but his storied service grew in World War II. He was at Guadalcanal in 1942, where he was wounded by a sniper and by shrapnel, earning his third Navy Cross. It was at Peleliu Island, in 1944, where he was the executive officer of the 7th Marines, that he earned his first Legion of Merit. In 1944, he earned his fourth Navy Cross for heroism at Cape Gloucester. He had now been in the Corps since 1918 and had served his nation well, but he was to have another opportunity when the Korean War began.

In 1950, at age fifty-two, he was in charge of the 1st Marine Regiment at the invasion of Inchon*, where he earned the Silver Star and his second Legion of Merit. He received his fifth Navy Cross and the Distinguished Service Cross for heroism at the Chosin Reservoir*. His unit was the rear guard, holding off the Chinese, while the rest of the Americans made it to safety.

He returned to command the 3rd Division at Camp Pendleton, California in 1952. He was promoted to major general and lieutenant general. In 1954, he commanded the 2nd Marine division. His health began to fail him, so he retired in 1955. As late as 1966, he requested duty in Vietnam, but was turned down because of his age. He died in October, 1971, at the age of seventy-three and is buried in Cemetery, Christchurch, Virginia, located just off General Puller Highway.

The legend of Chesty Puller lives on and he continues to be honored as "The Marines Marine." There is a Navy frigate named the USS Lewis Puller, a hall at Yorktown, Virginia, named Puller Hall, a postage stamp issued in 2005. One of his great honors is that the past, present, and future Marine Corps mascots, all English bulldogs, are named Chesty Pullerton.

He is remembered for one of his more famous statements, made at Chosin, during one of the darkest moments for the U.S. Marines: "Remember that you are the 1st Marines and not all the communists in hell can overrun you."

Reckless
U.S. Marine Corps

Reckless was a sergeant in the U.S. Marine Corps. Reckless was a horse and a member of the 1st Marine Division in Korea.

Reckless was owned by a young Korean jockey, in desperate need of money to get an artificial leg for his sister, who'd lost a leg from a land mine. Reckless was the only asset that he had of any value, so he reluctantly sold her to some Marines from a 1st Division Recoilless Rifle unit. They wanted Reckless to be an ammo bearer, but first they had to find out if she could handle the noise from their weapons and the noise at the front. Reckless was nervous at first but eventually learned to tolerate the noise. After her baptism of fire, she made many trips from the ammo dump up to the guns. After a time, she was promoted to sergeant and won an award for gallantry in action.

When her Marine unit left Korea, she was left behind, but Lieutenant Colonel Andrew Geer wrote several articles about Reckless and a book, *Reckless: Pride of the Marines*. Colonel Geer wrote of her, "Every yard she advanced was showered with explosives. Fifty-one times she marched through the fiery gantlet (sic) of the Red barrage—and she saved the day for the Leathernecks." His articles encouraged his readers to raise the money to pay for her passage to the States. It was agreed that Reckless should be stationed at Camp Pendleton, California.

Reckless was at Camp Pendleton for five years, when she was promoted to staff sergeant. Reckless died in 1968, leaving three offspring.

This story of Reckless is an excerpt from a November 14, 2004 article written by Lillian Cox for the San Diego Union-Tribune and reprinted with permission.

Matthew B. Ridgeway
U.S. Army

General Omar Bradley said that when General Ridgeway changed the course of the Korean War, "It was the greatest feat of personal leadership in the history of the Army." In a career that had many triumphs, the Korean War was General Ridgeway's greatest.

Matthew Bunker Ridgeway was born into a military family in 1895. His father had been a colonel in the artillery and Matthew went to West Point [U.S. Military Academy] to satisfy his father. He graduated in 1917 as a second lieutenant, but failed to get into World War I. Instead he sat out the war as an instructor at West Point. He was sent to China to be a company commander in the 15th Infantry. In 1927, he was sent to Nicaragua to monitor the elections. In 1930, he went to the Philippines, 1935 to Command and General Staff School, and 1937 to the Army War College. He then became Deputy Chief of Staff, 2nd Army, Assistant Chief of Staff of the 4th Army, and in 1942, he finally made brigadier general. In August, 1942, he became a major general and was put in command of the 82nd Airborne, which took part in the invasion of Sicily in 1943.

General Ridgeway jumped with his men during Operation Overlord, the invasion of Normandy, France, on June 6, 1944. His next exploit came when he was in charge of the American airborne troops in "Operation Varsity" in March, 1945. Sixteen thousand Allied paratroopers jumped behind the German lines, capturing Rhine River bridges and towns, to aid the advancing Allied armies. It was the largest airborne operation in WWII. In June, 1945, he was promoted to lieutenant general.

In December, 1950, he was given command of the 8th Army in Korea. The 8th and United Nations forces were still in retreat, from battles at the Chosin Reservoir* and the Ch'ongch'on River*, when he took command. He replaced division commanders who had served six months and worked to restore confidence in his army. He stopped the Chinese advance and led a counter-offensive in 1951.

In April, 1951, Ridgeway replaced General MacArthur as commander of the UN forces in Korea and as military governor of Japan. In May, 1952, Ridgeway replaced General Eisenhower as Supreme Allied Commander of NATO forces in Europe. In 1953, his final and highest level position in his military career was as Chief of Staff of the U.S. Army. In June, 1955, after thirty-eight years in the U.S. Army, he retired. In 1993, he died in Fox Chapel, Pennsylvania.

In a long and distinguished career, Matthew Ridgeway's masterful handling of the UN forces, in Korea, when he stopped the Chinese, stands out as his crowning achievement.

Syngman Rhee
President of the Republic of Korea

Syngman Rhee had, at times, been called corrupt, an embezzler, a dictator, and a murderer, but he was an anti-communist and the duly elected first president of the Republic of Korea and an ally of the United States.

Syngman Rhee was born March 26, 1875, to an aristocratic Korean family. As a young man, he was active in the struggle against Japan, and was arrested in 1897, at an anti-government demonstration, but not released until 1904. He went to the United States, where he attended George Washington and Harvard universities and earned a Ph.D. at Princeton University. He returned to Korea in 1910, the same year that Japan took control of the country. In 1919, he was elected president of the Provisional Government of the Republic of Korea, a position he held for six years before being impeached for misuse of authority.

At the end of World War II, when the Japanese lost control of Korea, Rhee—backed by the U.S.—was appointed head of the Korean government. His campaign to rid the country of communists was a ruse to eliminate all political opponents. In August, 1948, following elections in the South, Rhee was named president of the Republic of Korea. He almost immediately assumed dictatorial powers, outlawed political dissent, and allowed the internal police force to detain, torture, and kill suspected Communists and North Korean agents.

The North Koreans invaded in June, 1950, and swiftly moved south. When it appeared that the North Koreans were going to take Seoul, Rhee told his countrymen he would stay and fight, but instead he fled Seoul and left the people to the North Korean occupiers. Rhee set up a temporary government in Pusan*, on the southern coast.

After the recapture of Seoul, Rhee fell out of favor with the U.S. During the series of peace negotiations, he refused proposal after proposal, wanting, instead, a total victory that would make him ruler of a united Korea and eliminate Kim Il-sung, the absolute ruler of North Korea.

Rhee's corruption and political repression made it probable that he would be defeated in the 1952 elections. When his attempt to have an amendment exempting him from the two-term presidential limit was rejected, Rhee had opposition politicians arrested and the amendment passed. He won reelection by seventy-four percent. In 1960, he won his fourth term with ninety percent of the vote, perhaps because his opponent died, or was killed, shortly before the election.

When his candidate for vice president appeared to win by a rigged vote, Rhee was forced to resign in disgrace in April, 1960. He and his family were moved to Honolulu, Hawaii, where he died July 19, 1965, at ninety years old.

USS Toledo

The USS Toledo, named for the city of Toledo, Ohio, was christened by Mrs. Edward J. Moan of Toledo, whose son, Lieutenant Commander Floyd E. Moan, was a hero in the Pacific Theater in World War II. The champagne used to christen the ship was from grapes grown on Middle Bass Island, near Toledo. The bottle was made in Toledo by the Owens-Illinois Glass Company.

The USS Toledo (CA-133), a heavy cruiser with a crew of 1,142 officers and men, was commissioned on October 27, 1946. After her shakedown cruise, she sailed on the first of four peacetime tours in the western Pacific, visiting ports in Japan and Korea, to support American occupation forces.

When the North Korean forces invaded South Korea, the Toledo sailed to the Pacific and her first combat duty. After stops in Pearl Harbor and Sasebo, Japan, she was positioned on the Korean east coast and, along with other task force ships, bombarded North Korean facilities and front line troops. She then moved north along the coast, bombing roads and bridges to disrupt North Korean army supply lines.

The USS Toledo's most important Korean War mission came during the invasion at Inchon*, on Korea's west coast. Wolmi-do was a heavily armed island in the harbor at Inchon that had to be destroyed before the troops could go ashore. When destroyers entered the harbor to draw fire from the island—which determined the position of the enemy guns—the Toledo and other cruisers then fired on those gun positions until the island was neutralized. Following the bombardment, the 5th Marines captured the island. The Toledo then supported the 1st Marines landing south of Inchon. The Toledo provided fire support until October, when the fighting moved inland beyond the range of her guns.

A second amphibious operation took place at Wonsan*, on the east coast of Korea. The Toledo again provided bombardment of the shore, prior to the landings, and continued bombarding the North Koreans as the Marines advanced inland. After a three-month overhaul in the States, the Toledo returned in April, 1951 for her second tour of duty in the Korean War. She sailed along the west coast of Korea, providing support to I Corps, and along the east coast near Wonsan. She continued to provide support until November when she returned to the States. In September, 1952, the Toledo went for her third and final combat tour to the Korean War zone. She again provided bombardment of enemy positions. In February, 1953, she returned to the States, her combat duty in Korea completed. She entered the Hunters Point Naval Shipyard in California for an overhaul and was there when the war ended with the armistice of July 27, 1953. She later returned to Korean waters, even though hostilities had ended, to patrol the Korean coastline.

In January, 1955, on her eighth deployment to Asia, the Toledo took part in an evacuation of Nationalist Chinese forces to Taiwan, but the ship would not engage in combat duty again. She continued to serve in the western Pacific until January, 1960, when she was decommissioned. In 1974, the USS Toledo was sold for scrap, but her nameplate was saved and is mounted on a wall of the Toledo Club in Toledo, Ohio.

This may seem to be an ignominious end to a ship that had earned five battle stars, but the decommissioning is a solemn ceremony. It is attended by the squadron or unit commander, other senior officers, and government officials. The National Anthem is played, a chaplain gives the invocation, and a speaker gives a resume of the ship's history. The order is then given to strike eight bells, the colors are hauled down, and the watch, no longer required, is ordered to be secured.

The crew member with the most years of service is given the ship's ensign and the chaplain gives the benediction to end the ceremony—a fitting ending for a proud ship of the United States Navy.

HISTORIES

The Two Koreas: Facts and Figures

North and South Korea occupy the Korean peninsula, an area of approximately 85,000 square miles. The peninsula is bordered on the north by China and on the east by Russia. The Yellow Sea is to the west, the Sea of Japan to the east. Japan lies to the south, separated by the Korean Straits.

The population of South Korea is about 49 million, of which 10.5 million live in Seoul, 3.6 million in Busan, and 2.5 million in Daegu. Fifty-two percent profess to be Christian or Buddhist, while forty-six percent have no religious preference. South Korea has a democratic form of government with a president, an assembly, and a judiciary. The CIA World Factbook describes South Korea as a "fully functioning modern democracy." Ninety-eight percent of South Koreans are literate and the country placed first in problem-solving, third in mathematics, and seventh in science in the 2006 Organization for Economic Co-operation and Development's Programme for International Student Assessment.

The population of North Korea is estimated at 24 million, of which 3.2 million live in the capital city of Pyongyang. Although the majority of North Koreans are identified as irreligious, they have a strong Buddhist and Confucian cultural heritage, along with more recent Christian movements. The country suffered intermittent famines through the 1990s; Japan and South Korea sent numerous food shipments, as did the United States from 1997-2004, by which time agricultural production had recovered somewhat. North Korea is an hereditary dictatorship, with a cult of personality focused on the country's founder, Kim Il-sung, and now his ailing son, Kim Jong-il. The leader of the single-party state holds all power, although there is a Supreme People's Assembly. The North Korean literacy rate is ninety-nine percent.

According to Korean medieval-era records, the Gojoseon kingdom in Northern Korea and Manchuria was founded in 2333 BCE; the Southern Korea Jin state was founded in the third century BCE. In the late ninth to early tenth centuries, the several peninsula kingdoms were united under the Goryeo Dynasty, from which the country took its name. Over the centuries, the Korean peninsula was the site of colliding interests, as both invaders and warring parties within created both internal and external pressures and conflict.

In 1911, Imperial Japan annexed the Korean peninsula, but various resistance movements flourished in Manchuria, China, and Siberia. At the end of World War II and Japan's surrender, the Americans divided the peninsula at the 38th parallel, splitting control between the United States and the Union of Soviet Socialist Republics. Free elections scheduled for 1948 did not take place in the North and a communist government was installed there instead. Cross-parallel skirmishes and raids continued over the next several years. Despite the June, 1950 United Nations Security Council's recognition of the Republic of Korea as the sole legal government, Northern forces invaded the ROK on June 25, 1950.

The war that followed was sanctioned by the United Nations but fought primarily by the U.S. against the Soviet and Chinese-aided North. The July 27, 1953 armistice, recognizing the military stalemate, established a demilitarized zone (DMZ) 2.5 miles wide. Thousands of American troops remained in South Korea for years; a 2008 U.S.-South Korea agreement set the number of remaining U.S. troops at 28,500. Reunification talks, which had been ongoing, if sporadic, since the 1990s, have been suspended since 2010 when North Korea sank a ROK warship and later fired artillery shells at Yeonpyeong Island, a South Korean island only 7.5 miles from the North Korean coastline. That year the North Koreans proposed replacing the armistice with a formal peace treaty, but coupled the offer with a demand that international sanctions be lifted before the treaty's drafting. The U.S. declined to lift any sanctions until the North, which has an ongoing nuclear development program, takes steps to dismantle the program.

So the two Koreas remain technically at war, facing each other across the world's most heavily militarized border, with Seoul, the ROK capital, a mere thirty-five miles from the North Korean army stationed along the DMZ.

Another Tyrant, Another War

It was finally over, over there. Bells rang out in the few churches that still stood in the war-torn French countryside. Men emerged from their muddy, rat-infested trenches to shake hands with an enemy who had tried to kill them, just minutes before. This was the "war to end all wars." Never again would millions of young men give their lives in a war that no one wanted and no one understood. It was the eleventh hour of the eleventh day of the eleventh month of 1918 when the armistice that would end World War I was officially signed.

The damage was cleaned up; the church steeples, where snipers and forward observers had once hidden, were rebuilt. Life returned to normal, but not for long. Just twenty-one years later, on September 1, 1939, German forces invaded Poland. By December 7, 1941, the world was once again at war. WWI had been fought with the newest in weaponry—the machine gun, the tank, and the aeroplane. WWII was fought with super bombers, capable of destroying whole cities in a single raid, and atom bombs that could kill 100,000 people at a time.

For the totalitarian nations that had planned to quickly dispatch their enemies, this was a war where mass production and technology would decide the outcome. They had the buzz bombs, the super tanks, submarines, and kamikaze planes. Their opponents had B29 bombers, a massive Pacific fleet, thousands of tanks, and the atom bomb. This time it took fifty million deaths to prove the futility of war. Was this finally "the war to end all wars"?

To assure that there would be no more wars after WWI, the League of Nations had been formed to mediate disputes and prevent armed conflict. After its failure to stop Italy from conquering Abbysinia and Hitler from contravening the Versailles Treaty, most realized that the League of Nations was a toothless tiger. There would be no more false hopes after WWII. The United Nations was formed to oversee the world and guarantee peace forever. Of course, lasting peace continued to be as elusive as ever it had been.

When WWII ended, allies became enemies and enemies became allies, proving that politics and war create strange bedfellows. Germany, once the scourge of Europe, almost immediately became our bulwark against the Russian Communists bent on European domination. Japan, the enemy we hated for their inhumanity, was now our ally as we fought to save the Pacific, too, from communist domination. China, where Americans fought and died to defend the country from Japan, was now the enemy.

Even before WWII ended, the battle lines were drawn. Stalin promised that free elections would be held in the Eastern European countries the Russians had liberated. He, of course, had no intention of keeping his word, proving a saying that became popular at that time, "It is better to be conquered by the Germans than to be liberated by the Russians." Mao Zedong forced Chiang Kai-shek out of mainland China to Taiwan and China became the largest communist country in the world, and sponsor of the fledgling Democratic People's Republic of North Korea.

So before we had fully recovered from WWII and before the Gold Star mothers had dried their tears, we were faced with another war of major proportions. Just five years after the end of one war, the next began.

No matter the cost in human lives, no matter the damage to their country and no matter that history records say that most will fail, something drives tyrants to wage war on their fellow man. In the last hundred years, we have had Hitler, Mussolini, Stalin, Mao, and Tojo; and in 1950, Kim Il-sung of North Korea appeared on the world stage. Like his predecessors, he was bent on conquest. He decided that since he had the support of Communist China and Communist Russia, it was time to act. On June 25, 1950, the Korean People's Army invaded South Korea. Another tyrant had begun yet another war.

Armed Services Integration

It is important to recognize in this telling of the Korean War that the armed services had been segregated until President Harry S. Truman ordered their desegregation in 1948. By issuing Executive Order 9981, the President could avoid Congress and its Southern members who would not support desegregation. As it was, the Army stalled, finally announcing plans to desegregate in July, 1951.

In WWII, the Army had been almost completely segregated. There were all-black units, like the Red Ball Express who distinguished themselves, hauling supplies around the clock from supply depots to the front lines. The Buffalo Soldiers, of the 92nd Infantry Division, had first been organized in 1866 and were called "The Negro Cavalry." They fought their way up the Italian peninsula against some of the best units in the German Army, and had two Medal of Honor recipients. The Tuskegee Airmen—so named because they were stationed in Tuskegee, Alabama, an unlikely place for such a unit to live and train—were an all-black fighter group who protected bombers on missions all over south and central Europe. They flew P51 fighters with red tails, which their enemies came to recognize and fear.

During the Korean War, I was in a fully-integrated, anti-aircraft artillery unit in Austria. Integration was a major issue to the Army brass, but to us GIs it was not, as we worked together, ate together, and slept in the same barracks. We had a black first sergeant who was a career Army man with a chest full of ribbons on his class A uniform. He had the respect of everyone in our unit, not only for his rank, but also for his long and devoted service to his country. His race meant nothing to us.

In October, 1951, the last all-black unit, the 24th Infantry Regiment, was disbanded and integrated into other units. By the end of the Korean War, 600,000 black Americans had served in the military and about 5,000 of that number died. Black Americans distinguished themselves in the fighting, along with their white comrades. Captain Daniel "Chappie" James, Jr., commander of a fighter squadron, was one of the more famous. He flew 101 missions and received the Distinguished Flying Cross. Private First Class William Thompson and Sergeant Cornelius Charlton were both awarded the Medal of Honor, posthumously.

Dissent: A Great American Tradition

. . . and were it left to me to decide whether we should have a government without
newspapers or newspapers without a government,
I should not hesitate to choose the latter.
— Thomas Jefferson

Since the First Amendment became law on December 15, 1791, the American press and the public have not hesitated to exercise their right to free speech, freedom of the press, and the right of assembly. The right to challenge the government is the most cherished of our freedoms. This has never been truer than in time of war or impending war. And what begins in the press often moves to the streets.

Before there was a First Amendment or even a United States of America, dissent was a tactic of the independence-minded colonists. Thomas Paine wrote the pamphlet "Common Sense," which inspired the colonists to fight for freedom from the British. When the British Parliament imposed more and more taxes on the colonies, the people reacted with the "Boston Tea Party," dumping a shipload of tea into Boston Harbor. On the other side, dissenters were called loyalists. They may have been in the minority, but they raised their voices in favor of continued rule by England.

During the War of 1812, some New York militias refused to respond when called to serve. In Baltimore, Maryland, there were anti-war riots and peace societies were formed to oppose the war. Pro-war demonstrators destroyed an anti-war newspaper office.

The Civil War draft met with strong opposition, particularly in New York City. In 1863, there were four days of draft riots; 500 people attacked a draft office. The people opposed the law that allowed a rich man to pay a $300 commutation fee to stay out of the Army. It was estimated that more than 1,000 people were killed and many thousands more were injured in the rioting. There was also violence in the Confederate states over the practice of the rich hiring substitutes and the draft's exempting the rich planter class.

In World War I, opposition to a foreign war came from the isolationists. Since the poor were thought to be more expendable, conscription was often determined by social class. There were some violent protests, some draft dodgers, and some desertions. Conscripts who refused to serve were jailed until war's end. After the war, isolationism was again on the rise. This was to have an effect on President Franklin Roosevelt's efforts to help the British in World War II.

The American public accepted WWII because it was thrust upon them. They accepted the draft of ten million men because it was conducted in a fairer manner than in past wars, with fewer men deferred. The America First Committee, an organization of 800,000 members, was led by such well-known public figures as Charles Lindbergh, North Dakota Senator Gerald Nye, and Michigan Senator Arthur Vandenberg. They traveled the country speaking out against America's entering the war. But on December 8, 1941, they ceased their efforts and advised FDR that they supported the war.

And what about the Korean War? This war certainly should have aroused the public's protest fervor. It was to be fought on foreign soil, our vital assets were not at risk, it was fewer than five years after the end of WWII, and it was a political war—not for a military victory, but for containment of communism. For all of these reasons and the American people's penchant for isolationism, protests might have been expected. But, more protests came after WWII and before the Korean War over joining the United Nations. The controversy brought back all the arguments raised after WWI against the League of Nations. As early as the post-Revolutionary era, George Washington had warned the young nation about the danger of "permanent alliances with the foreign world." After WWI, Congress had voted down joining the League of Nations to keep America free of involvement in the often petty European wars. The hope that the UN would prevent future wars and contain communism led the United States to join in 1945.

So despite a long history of organized protests against America's wars, the public generally accepted the need for the Korean War and the Selective Service Act of 1948. From 1950 to 1953, 1,500,000 men were inducted. A Gallup poll in 1953 revealed that seventy percent of Americans felt that the draft was fair. The communists were so aggressive in their post-war attempts to dominate Europe and Asia that the American public agreed with the UN's efforts of containment and thus the U.S. participation in the Korean War.

KOREAN WAR VETERANS ASSOCIATION

In July, 1985, William T. Norris, a former sergeant in Company F, 27th Infantry Regiment, 25th Infantry Division, founded the Korean War Veterans Association. He was one of forty original members who had met in Arlington, Virginia. Their objective was outlined in their mission statement:

DEFEND our Nation

CARE for our Veterans

PERPETUATE our Legacy

REMEMBER our Missing and Fallen

MAINTAIN our Memorial

SUPPORT a free Korea

The organization of those who had served honorably in the U.S. military between June 25, 1950 and January 31, 1955 would provide a means for members to contact other members, to establish memorials for those who served in the Korean War, to aid needy members and/or their families, to maintain a national headquarters, and to grant charters to local or regional organizations. Among their intentions was to refute the Korean War's title, "The Forgotten War."

Northwest Ohio KWVA #13 was chartered in 1996 with twenty-three members. Today the chapter meets monthly in Toledo, Ohio; membership is approximately 110. The members are a classic example of community outreach, cooperating with local government and schools. Every year, they meet with schoolchildren to inform them of the Korean War. They know that the children are interested because the veterans are invited back every year.

Downtown Toledo's Civic Center Mall had a World War II memorial and a Vietnam memorial, so it seemed fitting that a Korean War memorial be placed there as well. A number of meetings were held with the Toledo Arts Council, who set the standard for the memorial. Fundraising was a cooperative effort with the help of then-Toledo Mayor Jack Ford, Lucas County, the Lucas County veterans group, and KWVA #131. Through their combined persistence, the Korean War Memorial finally took its place of honor on the mall.

In their mission statement twenty-five years ago, William T. Norris and his charter members set the goals for Korean War Veterans Associations everywhere. Northwest Ohio KWVA #131, a proactive organization, consistently strives to meet those goals.

KOREAN WAR VETERANS
HISTORY PROJECT LIST

List of all veterans history project interviews of Korean War veterans in the Ward M. Canaday Center of The University of Toledo's Carlson Library (as of December 2011).

Karl Alberti
Branch: Army Air Corps;
Year: 1942-1945;
Battalion-Regiment: 8th/9th Air Force;
Highest Rank: Master SergeantWWII,
1st Lieutenant;
Conflict: World War II, Korea

David D. Antonacci
Branch: Navy;
Year: 1952-1960;
Battalion-Regiment: U.S.S. Floyds Bay
AVP 40;
Highest Rank: Petty Officer 3rd Class;
Conflict: Korea

Elmer J. Balough
Branch: Air Force;
Year: 1947-1950;
Battalion-Regiment: 5th Air Force;
Highest Rank: Sergeant;
Conflict: Korea

Leo D. Barlow
Branch: Marine Corps;
Year: 1950-1954;
Battalion-Regiment: 1st Company,
3rd BTN 1st Regiment 1st Marine Division;
Highest Rank: Sergeant;
Conflict: Korea

Frank J. Bartell
Branch: Navy;
Year: 1942-1953;
Highest Rank: Lieutenant Junior Grade;
*Conflict:*World War II, Korea

Vernon J. Basilius
Branch: Navy;
Year: 1952-1954;
Battalion-Regiment: U.S.S. Hunt DD-674;
Highest Rank: Petty Officer 2nd Class;
Conflict: Korea

Gilbert Berry
Branch: Army;
Year: 1947-1951;
Battalion-Regiment: 187th RCT (Airborne);
Highest Rank: Corporal Reserve E-7;
Conflict: Korea

William D. Birr
Branch: Navy;
Year: 1954-1956;
Battalion-Regiment: Neville Island
Ordinance Depot;
Highest Rank: Corporal;
Conflict: Korea

Donald Callahan
Branch: Air Force;
Year: 1950-1953;
Battalion-Regiment: 19th Bomb Group,
30th Bomb Squadron;
Highest Rank: Airman 1st Class;
Conflict: Korea

Karlene A. Carpenter
Branch: WAVES;
Year: 1944-1952;
Battalion-Regiment: Hospital Corps;
Highest Rank: Pharmacist Mate 2nd Class;
Conflict: World War II, Korea

Joseph Carter
Branch: Army;
Year: 1945-1951;
Battalion-Regiment: 4th Army Division,
10th Corp. Korea;
Highest Rank: Corporal;
Conflict: World War II, Korea

William Cartledge
Branch: Marine Corps;
Year: 1948-1952;
Battalion-Regiment: Baker Co,
1st BTN 7 Re 6t 1st Division;
Highest Rank: Corporal;
Conflict: Korea

Walter A. Churchill
Branch: Marine Corps;
Year: 1947-1977;
Battalion-Regiment: C Company
1st BN 7th Regiment;
Highest Rank: E9 Master Gunnery;
Conflict: Korea

Harry W. Clark
Branch: Army;
Year: 1944-1966;
Battalion-Regiment: 8th Infantry Division;
Highest Rank: E8 Master Sergeant;
Conflict: World War II, Korea

William H. Clegg
Branch: Navy;
Year: 1944-1953;
Highest Rank: Captain;
Conflict: World War II, Korea

Thomas A. Cowher
Branch: Navy;
Year: 1950-1952;
Battalion-Regiment: Hospital Corpsman;
Highest Rank: Hospitalman;
Conflict: Korea

Kenneth E. Cox
Branch: Marine Corps;
Year: 1948-1952;
Battalion-Regiment: 1st Marine Regiment,
1st Marine Division;
Highest Rank: Sergeant;
Conflict: Korea

Charles L. Cromly
Branch: Navy;
Year: 1942-1945;
Battalion-Regiment: US Mobile Hospital
#5, 7;
Highest Rank: Lieutenant Senior Grade;
Conflict: World War II, Korea

Robert Darr
Branch: Marine Corps;
Year: 1950-1954;
Battalion-Regiment: 5th Regiment 1st BN,
Motor Transport C;
Highest Rank: Sergeant E4;
Conflict: Korea

James M. Dashbach
Branch: Air Force;
Year: 1955-1957;
Battalion-Regiment: 7331st Technical
Training Wing;
Highest Rank: Lieutenant Colonel;
Conflict: Korea

Joel C. Davis
Branch: Army;
*Year:*1951-1953;
Battalion-Regiment: 772 MP BN;
Highest Rank: Corporal;
Conflict: Korea

Herman R. DeBrosse
Branch: Army;
Year: 1950-1952;
Battalion-Regiment: 24th Division,
21st Regiment;
Highest Rank: Corporal;
Conflict: Korea

Milo H. Downs Jr.
Branch: Navy;
Year: 1940-1975;
Battalion-Regiment: 3rd, 5th, 6th Fleet;
Highest Rank: Master Chief Petty Officer;
Conflict: World War II, Korea, Vietnam

Richard D. Drzewiecki
Branch: Navy Reserves;
Year: 1951-1953;
Battalion-Regiment: Seabees
(construction BTN);
Highest Rank: Construction Electrician;
Conflict: Korea

Richard H. Eckert
Branch: Navy;
Year: 1946-1976;
Battalion-Regiment: U.S.S. Lexington,
Patrol Squadron 42-46-1;
Highest Rank: Captain;
Conflict: Vietnam, Korea

Paul L. Erdy
Branch: Air Force;
Year: 1952-1956;
Battalion-Regiment: Far East Air Force
Command 6000th S;
Highest Rank: A1P;
Conflict: Korea

Eugene F. Eversole
Branch: Army;
Year: 1943-1971;
Battalion-Regiment: Artillery;
Highest Rank: Captain;
Conflict: World War II, Korea, Vietnam

Robert E. Fuller
Branch: Army;
Year: 1951-1954;
Battalion-Regiment: 2nd Infantry Division
23 Regiment;
Highest Rank: PFC;
Conflict: Korea

Gary E. Gabriel
Branch: Air Force;
Year: 1950-1953;
Battalion-Regiment: Office of
Special Investigations (OSI);
Highest Rank: 1st Lieutenant;
Conflict: Korea

Novarro Gibson
Branch: Air Force/National Guard;
Year: 1952, 55, 68;
Battalion-Regiment: 19th Comm. Coast
Squadron 180th ANG;
Highest Rank: MSGT Master Sergeant;
Conflict: Korea, Vietnam, Desert Storm

John G. Griffin
Branch: Army;
Year: 1954-1956;
Battalion-Regiment: 29th Tank Battalion
HQ Platoon;
Highest Rank: SP3;
Conflict: Korea

Donald M. Griffith
Branch: Marine Corps;
Year: 1945-1954;
Battalion-Regiment: F Co, 2nd BN,
5th Regiment, 1st Marine Division;
Highest Rank: Staff Sergeant;
Conflict: Korea

Lawrence W. Grimm
Branch: Navy; *Year:* 1943-1951, 1946;
Battalion-Regiment: Hospital Corps;
Highest Rank: Pharmacist Mate 2nd Class;
Conflict: World War II, Korea

Fred Haddad
Branch: Army, Navy, Air Force;
Year: 1945-1954;
Battalion-Regiment: Army, Navy Reserve,
Air Force;
Highest Rank: Corporal, Seaman, Sergeant;
Conflict: World War II, Korea

Clenastine Hamilton
Branch: WACS;
Year: 1951-1954;
Battalion-Regiment: HQ DET STA COM
8030th AU;
Highest Rank: Corporal;
Conflict: Korea

Robert F. Hassen
Branch: Marines;
Year: 1948-1952;
Battalion-Regiment: 1st Marine Division
Ordinance BN;
Highest Rank: Corporal;
Conflict: Korea

James L. Hays
Branch: Marine Corps;
Year: 1949-1953;
Battalion-Regiment: 2nd Battalion
1st Marines, 1st Division;
Highest Rank: Sergeant;
Conflict: Korea

Joseph H. Heiny
Branch: Army;
Year: 1950-1952;
Battalion-Regiment: 1st Cavalry,
7th CAV Regiment HQ&HQ Co.;
Highest Rank: Staff Sergeant;
Conflict: Korea

Ernest R. Heller
Branch: Navy;
Year: 1950-1954;
Battalion-Regiment: U.S.S. Columbus CA74;
Highest Rank: Machine Repairman 1st Class
MRI;
Conflict: Korea

Richard Henning Jr.
Branch: Marine Corps;
Year: 1952-1954;
Battalion-Regiment: 1st & 5th
Marine Division;
Highest Rank: Corporal;
Conflict: Korea

Lloyd Hill
Branch: Army;
Year: 1951-1953;
Battalion-Regiment: Army Chemical Corps;
Highest Rank: Private First Class;
Conflict: Korea

John W. Hinds
Branch: Army;
Year: 1948-1953;
Battalion-Regiment: 235th Field Artillery
Observation BN;
Highest Rank: 1st Lieutenant;
Conflict: Korea

Taylor Horne
Branch: Navy;
Year: 1941-1954;
Battalion-Regiment: Motor Torpedo Boat
Squadron-Mil. Air Trans;
Highest Rank: Staff Sergeant;
Conflict: World War II, Korea

Earl R. Hufford
Branch: Army;
Year: 1951-1953;
Battalion-Regiment: Medical Corps
11th Evacuation Hospital;
Highest Rank: Private 1st Class;
Conflict: Korea

Reginald S. Jackson
Branch: Army;
Battalion-Regiment: 37th Division;
Highest Rank: Lieutenant Colonel;
Conflict: World War II, Korea

Richard A. Janicki
Branch: Army;
Year: 1951-1953;
Battalion-Regiment: 630th Light Equipment
Combat Eng. Co.;
Highest Rank: Private 1st Class
Conflict: Korea

Robert B. Juergens
Branch: Navy;
Year: 1943-1946;
Battalion-Regiment: 16th NCB, 126th NCB;
Highest Rank: Lieutenant Senior Grade;
Conflict: World War II, Korea

Ralph Keefe
Branch: Navy;
Year: 1947-1951;
Battalion-Regiment: U.S.S.
Gen. Geo Randall;
Highest Rank: Electricians Mate 3rd Class;
Conflict: Korea

Laurence E. Kish
Branch: Army;
Year: 1952-1954;
Battalion-Regiment: 40th Infantry Division
223 Infantry Regiment;
Highest Rank: Master Sergeant;
Conflict: Korea

Dale H. Kuhlman
Branch: Navy;
Year: 1942-1946; 1951-1953;
Battalion-Regiment: Armed Guard Officer;
Highest Rank: Commander;
Conflict: World War II, Korea

Edward J. Kusina
Branch: Marine Corps;
Year: 1949-1952;
Battalion-Regiment: 1st Marine Division;
Highest Rank: Sergeant;
Conflict: Korea

Wendell Lawrence
Branch: Air Force;
Year: 1951-1976;
Battalion-Regiment: Tactical Air Force;
Highest Rank: Colonel;
Conflict: Korea, Vietnam

Robert A. Lempke
Branch: Army;
Battalion-Regiment: Love Co. 23rd Infantry
Regiment, 2nd Infantry Division;
Highest Rank: Sergeant 1st Class;
Conflict: Korea

Glenn E. Levy
Branch: Navy & Coast Guard;
Year: 1943-1966;
Battalion-Regiment: AM101 Harold
Minesweeper - APA179;
Highest Rank: Chief Electrician E7;
Conflict: World War II, Korea, Vietnam

David Ludwikoski
Branch: Army;
Year: 1944-1946; 1950-1953;
Battalion-Regiment: 96th Division, 382
Regiment (WW II);
Highest Rank: Tech. Sergeant;
Conflict: World War II, Korea

John F. MacDonald
Branch: Navy;
Year: 1948-1983;
Highest Rank: Captain (O6);
Conflict: Korea, Vietnam

Dean McPherson
Branch: Air Force;
Year: 1952-1956;
Highest Rank: Airman 1st Class;
Conflict: Korea

Ross Mergenthaler
Branch: Army Air Force;
Year: 1950-1953;
Battalion-Regiment: WWII, 20th Air Force,
Korea 433rd Tr;
Highest Rank: Staff Sergeant;
Conflict: World War II, Korea

Francis Mick
Branch: Navy;
Year: 1948-1952;
Battalion-Regiment: U.S. Naval Hospital,
Chelsea, MA;
Highest Rank: 2nd Class Petty Officer;
Conflict: Korea

Donald Millington
Branch: Marine Corps;
Year: 1951-1953;
Battalion-Regiment: 1st Marine Division 1st
Regiment Fox Co.;
Highest Rank: Sergeant;
Conflict: Korea

Terence Mohler
Branch: Army;
Year: 1951-1953;
Battalion-Regiment: 623rd Artillery
10th Corps;
Highest Rank: Corporal;
Conflict: Korea

Don J. Mooney
Branch: Marine Corps;
Year: 1948-1951;
Battalion-Regiment: 1 Bn 7th Regiment
1st Division;
Highest Rank: Corporal;
Conflict: Korea

Donald Noethen
Branch: Navy;
Year: 1951-1955;
Battalion-Regiment: APA 152;
Highest Rank: 2nd Class Gunner's Mate;
Conflict: Korea

Robert E. O'Keefe
Branch: Marine Corps;
Year: 1951-1952;
Battalion-Regiment: Baker Co. 7th Marines;
Highest Rank: Corporal;
Conflict: Korea

Frederick E. Peppers
Branch: Navy;
Year: 1936-1957;
Battalion-Regiment: U.S.S. West Virginia,
Bittern & Finch;
Highest Rank: Signal Man Chief;
Conflict: World War II, Korea

Richard Piriczky
Branch: Air Force;
Year: 1949-1959;
Battalion-Regiment: 3rd Bomb Wing,
18th Fighter Bomber;
Highest Rank: Corporal Airman 2nd Class;
Conflict: Korea

Charles W. Poore
Branch: Navy Reserves;
Year: 1945-1987;
Battalion-Regiment: VR732 Squadron;
Highest Rank: Aviation Machinists 1st Class;
Conflict: World War II, Korea, Vietnam

Lyman S. Prater
Branch: Navy;
Year: 1942-1954;
Highest Rank: Lieutenant Commander;
Conflict: World War II, Korea

Robert Primisch
Branch: Navy/Air Force;
Year: 1945, 46, 56;
Battalion-Regiment: Sub Chason PC
579ARS 29 Salvage Ship;
Highest Rank: S/Sergeant;
Conflict: World War II, Korea, Vietnam

Donald Proudfoot
Branch: Army;
Year: 1951-1953;
Battalion-Regiment: 40th Division 140th
Tank BN;
Highest Rank: Corporal;
Conflict: Korea

Richard Rajner
Branch: Army;
Year: 1966-1971;
Battalion-Regiment: 76th Eng. Bn, 610th
MTCE Bn, HQ & HQ;
Highest Rank: Sergeant T5;
Conflict: Korea, Vietnam

Alfred Reiser
Branch: Army;
Year: 1950-1952;
Battalion-Regiment: 5th Regimental Combat Team, 24th Division;
Highest Rank: Corporal;
Conflict: Korea

John Repp
Branch: Air Force;
Year: 1944, 46, 50;
Battalion-Regiment: 2nd Emergency Squadron; *Highest Rank:* S/Sergeant;
Conflict: World War II, Korea

George M. Rinkowski (Rev.)
Branch: Army Air Force;
Year: 1945-1956;
Highest Rank: Major;
Conflict: World War II, Korea

Marion A. Risk
Branch: Navy & Marine Corps;
Year: 1942-1955;
Battalion-Regiment: Naval Av. Cadet, Marine Air Group 23;
Highest Rank: Captain;
Conflict: World War II, Korea

Harry P. Schulman
Branch: Army;
Year: 1946-48, 51;
Battalion-Regiment: Supply Team 98 Qt. Master SVC, I Co;
Highest Rank: PFC;
Conflict: Korea

Walter W. Schumacher
Branch: Navy;
Year: 1943-1947;
Battalion-Regiment: Light Cruiser U.S.S. Omaha;
Highest Rank: 2nd Class Machinist;
Conflict: World War II, Korea

Donald E. Schwarz
Branch: Navy;
Year: 1941-1971;
Battalion-Regiment: DD 414 U.S.S. Russell, U.S.S. Rudler;
Highest Rank: Cox. USGQ CWO (Ret.);
Conflict: World War II, Korea, Vietnam

Daniel C. Seemann
Branch: Marine Air Force;
Year: 1952-1983;
Battalion-Regiment: Marine Air Wing;
Highest Rank: Colonel;
Conflict: Korea, Vietnam

Robert E. Shake
Branch: Army;
Year: 1942-1964;
Battalion-Regiment: 406th Anti-aircraft Battery D;
Highest Rank: Sergeant 1st Class;
Conflict: World War II, Korea

Lester Sharrit
Branch: Army, Navy, Marine;
Year: 1940-1953;
Battalion-Regiment: Navy, U.S.S. Nassau, 1st Marine Division;
Highest Rank: Torpedoman 1st Class;
Conflict: World War II, Korea

Gordon C. Sloan
Branch: Navy;
Year: 1944-1946;
Battalion-Regiment: 3rd Naval Fleet;
Highest Rank: Seaman 1st Class;
Conflict: World War II

Gordon W. Sloan
Branch: Army;
Year: 1954-1955;
Highest Rank: Intelligence Specialist;
Conflict: Korea

Paul D. Smith
Branch: Marine Corps;
Year: 1951-1953;
Battalion-Regiment: 2nd BN 7th Regiment
1st Marine Division;
Highest Rank: Private 1st Class;
Conflict: Korea

Robert E. Smithers
Branch: Marine Corps;
Year: 1952-1954;
Battalion-Regiment: 5th Marine Division
H Co 3rd BN;
Highest Rank: Private 1st Class;
Conflict: Korea

Robert J. Sollmer
Branch: Army & Marine Corps Reserve;
Year: 1945-1948;
Battalion-Regiment: 6th Constabulary Unit,
82nd Airborne;
Highest Rank: Corporal;
Conflict: Korea

David L. Souder
Branch: Army Engineers;
Year: 1958-1961;
Battalion-Regiment: 76th Eng.
Battalion Const.;
Highest Rank: Specialist 4th Class;
Conflict: Korea

Robert Swain
Branch: Army; *Year:* 1949-1951;
Battalion-Regiment: 1st Cavalry 8th
Regiment; *Highest Rank:* Private 1st Class;
Conflict: Korea

Donald Swain
Branch: Army; *Year:* 1951-1953;
Battalion-Regiment: 1st Cavalry,
8th Regiment L Company;
Highest Rank: Corporal;
Conflict: Korea

Harold Swicegood
Branch: Army;
Year: 1946-1970;
Battalion-Regiment: Airborne Rangers, Green
Beret;
Highest Rank: Sergeant 1st Class;
Conflict: World War II, Korea, Vietnam

Richard S. Thompson
Branch: Marine Corps;
Year: 1951-1952;
Battalion-Regiment: 1st Marine Division;
Highest Rank: PFC;
Conflict: Korea

Kathryn Todd
Branch: Woman's Air Force (WAF);
Year: 1952-1955;
Battalion-Regiment: 4402 Photography Unit;
Highest Rank: Staff Sergeant;
Conflict: Korea

Thomas J. Van Buren
Branch: Army Air Force;
Year: 1951-1987;
Battalion-Regiment: 7th Infantry Division,
Air National Guard;
Highest Rank: Guard Master Sergeant (ET);
Conflict: Korea

Richard Wagner
Branch: Army;
Year: 1949-1952;
Battalion-Regiment: 24th Division,
21st Regiment;
Highest Rank: Corporal;
Conflict: Korea

Frank O. Walbolt
Branch: Navy;
Year: 1951-1955;
Highest Rank: ME 2 Metalsmith;
Conflict: Korea

Phyllis E. Wells
Branch: Navy;
Year: 1953-1954;
Highest Rank: Navy Seaman;
Conflict: Korea

Clem Whittebort
Branch: Army;
Year: 1940-1954;
Battalion-Regiment: 37th Division
148th Infantry Regiment;
Highest Rank: PFC;
Conflict: World War II, Korea

William R. Williams
Branch: Navy;
Year: 1951-1954;
Battalion-Regiment: Wisconsin-Roanoke-New Jersey;
Highest Rank: Radar Man 3rd Class;
Conflict: Korea

Norman H. Yager
Branch: Marine Corps;
Year: 1952-1957;
Battalion-Regiment: Airing;
Highest Rank: S/Sergeant;
Conflict: Korea